SpringerBriefs in Energy

Energy Analysis

Series Editor

Charles A. S. Hall, Professor Emeritus, SUNY College of Environmental Science
and Forestry, Syracuse, USA

The "energy crises" of the 1970s, together with the appearance of numerous books and papers on the general theme of "limits to growth," catapulted energy from obscurity to social, economic, and academic prominence. Then, fuel prices came down again, economies recovered, and energy more or less disappeared from university and political discourse until about 2005. Many believed that market processes had resolved, and would continue to address, all energy supply issues. Now, energy in all of its aspects is back with a vengeance. While oil spills, mine deaths, and stock market plunges grab the headlines, there are much more fundamental discussions taking place about energy and its economic and environmental effects, many already occurring, on a societal scale. Most fundamental in this respect is the arrival, plus or minus a few years, of global peak oil, which in 1968 was first predicted to occur around the turn of the millennium. Clearly, energy never really went away and in fact underlies all physical motion, life, chemistry, and economics. It has been neglected in our understanding and teaching of societal processes for far too long, and particularly now when the scope and importance of issues growing out of energy availability and use are increasing yearly. These issues and impacts include the potential for economic growth and wealth creation (including so-called sustainable development), climate change, general pollution, agricultural production, clean water, and perhaps even the continued existence of civilization as we know it in the coming decades.

We believe that there is a great need for a comprehensive and integrated series of books that provides a quantitative accounting of energy use, including the potential and limitations of real-world deployments of new technologies. One key concept is Energy Returned on Energy Invested (EROEI or EROI), which is used to quantify the actual net energy for society from sources as diverse as "renewable" energy from solar and wind power installations to the current enthusiasm over unconventional oil and gas. **Springer Briefs in Energy Analysis** will cover the fundamental ways in which energy operates in the natural world and, in its abundance and governing physical laws, enables and constrains all human activities. The series will be empirically based with much technical information while remaining accessible to the non-specialist reader. Individual books will minimize the jargon, mathematics, or theory of specialist books in the various disciplines on the one hand, and the advocacy positions of popular accounts that may make their scholarship and conclusions suspect on the other.

Under the editorship of Charles Hall and a panel of energy experts, the series will include contributions from many of the world's most authoritative energy analysts. The series will appeal to anyone interested in energy use and its impacts on the economy and society in general.

The audience will range from advanced undergraduates through professionals in the physical science, environmental science, and economics and financial communities. Financial analysts will gain an understanding of how energy is increasingly impacting economic processes. University instructors will find these books to be invaluable for providing students (and themselves) with greater depth and insight into the role of energy in society with an emphasis on the methods and applications of energy accounting.

Timothy McWhirter

Maximum Power and its Philosophical Roots

The Critical Importance Today of the Ideas
of Howard Odum and Friedrich Nietzsche

 Springer

Timothy McWhirter
World Languages and Philosophy
Montgomery College
Rockville, MD, USA

ISSN 2191-5520   ISSN 2191-5539   (electronic)
SpringerBriefs in Energy
ISSN 2191-7876   ISSN 2199-9147   (electronic)
Energy Analysis
ISBN 978-3-031-80621-6   ISBN 978-3-031-80622-3   (eBook)
https://doi.org/10.1007/978-3-031-80622-3

© The Editor(s) (if applicable) and The Author(s), under exclusive license to Springer Nature
Switzerland AG 2025
This work is subject to copyright. All rights are solely and exclusively licensed by the Publisher, whether
the whole or part of the material is concerned, specifically the rights of translation, reprinting, reuse of
illustrations, recitation, broadcasting, reproduction on microfilms or in any other physical way, and
transmission or information storage and retrieval, electronic adaptation, computer software, or by similar
or dissimilar methodology now known or hereafter developed.
The use of general descriptive names, registered names, trademarks, service marks, etc. in this publication
does not imply, even in the absence of a specific statement, that such names are exempt from the relevant
protective laws and regulations and therefore free for general use.
The publisher, the authors and the editors are safe to assume that the advice and information in this book
are believed to be true and accurate at the date of publication. Neither the publisher nor the authors or the
editors give a warranty, expressed or implied, with respect to the material contained herein or for any
errors or omissions that may have been made. The publisher remains neutral with regard to jurisdictional
claims in published maps and institutional affiliations.

This Springer imprint is published by the registered company Springer Nature Switzerland AG
The registered company address is: Gewerbestrasse 11, 6330 Cham, Switzerland

If disposing of this product, please recycle the paper.

This book is dedicated to my partner in life, Nancy E. Kendrick.

Preface

H. T. Odum (1924–2002) was a prolific and influential American systems ecologist who published more than ten books and hundreds of scientific papers. A number of his papers helped create new fields of study, including ecological modeling, ecological engineering, and ecological economics. A number of his books were translated into other languages, including Spanish, Japanese, Chinese, and Russian, among others. He helped develop a thermodynamic approach to evolutionary theory and argued that natural selection was guided by what he called the *maximum power principle*. He analyzed all kinds of systems using this principle: physical systems, electrical systems, and ecological systems. He also used this principle to analyze human social systems and the moral and religious values associated with them.

The German philosopher Friedrich Nietzsche (1844–1900) developed a controversial principle he called the *will to power*, which he argued guided evolutionary and social development. He used this principle in his critique of morality, philosophy, religion, art, and culture. After his death, his writings went on to have a profound impact on philosophy, first in Germany and France, and then eventually in the United States and around the world. His work contributed to the development of existentialism, postmodernism, and psychoanalysis, and influenced writers such as W. B. Yeats and W. H. Auden, and composers such as Gustav Mahler and Richard Strauss.

In 2012, I published a paper on Nietzsche's critique of morality that briefly described how Nietzsche's concept of the will to power appeared to be similar, in important respects, to the maximum power principle (MPP) that was used by H. T. Odum. A decade later, I received an email from the ecologist Charles A. S. Hall, who was astonished that he was quoted by a philosopher regarding the principle. I later learned that this is a common practice for Charlie: He likes to contact different people who work in different disciplines and discuss ideas he is interested in from different perspectives. I told Charlie that philosophers generally viewed the version of the will to power that applies to both abiotic and biotic systems as being "empirically implausible." I investigated this issue in my dissertation and have continued to focus on it throughout my career as a philosopher. I suggested to Charlie that an in-depth investigation of the relation between the will to power and the MPP could have an impact on this situation. I also knew that if I had a chance to work with Charlie on this investigation, it would be a particularly unique and, for me, exciting opportunity, given the fact that he had worked with H. T. Odum as a Ph.D. student and, among his numerous publications, he had both written a paper entitled "The Continuing Importance of Maximum Power" (2004) and edited the book *Maximum Power: The Ideas and Applications of H. T. Odum* (1995). When Charlie agreed to work with me on this project, the idea for this book was born.

Then, Charlie told me to read Odum's book *Environment, Power, and Society* (1971), and this time it was I who was astonished. This book is amazing for many reasons, but one thing in particular jumped off the page for me: Odum critically analyzed moral and religious values from the perspective of the MPP. In my 2012 paper, I had argued that Nietzsche critically analyzed moral and religious values from a scientific perspective based on the will to power, understood as an empirical principle. When I saw this parallel, I knew that a book outlining the parallels and differences between Nietzsche and Odum had to be written and I was extremely fortunate to work with Charlie on it. (This book is my responsibility but Charlie has served as a comprehensive editor, as have two others.)

This book should be of interest to scientists from many different scientific disciplines, to philosophers with an interest in contemporary philosophy, philosophical naturalism, and Nietzsche, and to general readers with an interest in science and philosophy and the relation between them. Both scientists and philosophers should be interested in the empirical evidence that has emerged over the past 15 years that provides support for the MPP and the will to power (e.g., Lenton et al., 2016; Steffen et al., 2015; Nagel, 2012). While those authors do not mention these principles, their work nevertheless supports them. Many scientists and philosophers are not aware of how the MPP has been used successfully to describe phenomena in several different scientific disciplines. This book helps tell this story.

This investigation also illustrates how the MPP has been misinterpreted in contemporary scientific discussions, which should be of particular interest to scientists. Odum wrote that the optimum efficiency for maximizing useful power output may increase as the available energy decreases (Odum and Pinkerton, 1955; Odum, 2001, 2007). This important point is recognized by ecologists who worked with Odum, such as Mark Brown (2023), but its significance is overlooked by some contemporary scientists (Sciubba, 2011; Glazier, 2024).

This book should be of interest to philosophers for a number of reasons. First, this book will demonstrate that Nietzsche's version of the concept of the will to power that applies to biotic and abiotic systems is empirically plausible according to the sciences in the twentieth and twenty-first centuries, notwithstanding the views of philosophers who have suggested otherwise. Second, I develop a unique naturalistic interpretation of Nietzsche's critique of morality. I argue that Odum's analysis of moral and religious values represents a twentieth-century model of Nietzsche's critique of morality that provides us with a unique opportunity to better understand the nature of the perspective they share and the extent to which they remain consistent with it. Nietzsche and Odum both describe life and moral values evolving over time in a manner that is guided by, respectively, the will to power or the MPP, but

Odum appears to have more confidence in this thesis and develops it more explicitly than Nietzsche. This investigation illustrates that their work provides a unique view of morality and how it evolves over time, which should provide a valuable contribution to the contemporary debate in ethical theory, and could have broader implications for our understanding of the practice of philosophy.

This book should also be of interest to general readers with an interest in science and/or philosophy. The chapters are written so that they are accessible to readers from other disciplines and this should make them accessible to general readers. The books that Odum wrote for a general audience—*Environment, Power, and Society* (1971), *Energy Basis for Man and Nature* (1976), and *A Prosperous Way Down* (2001)—had a large influence on this book. Odum knew he was dealing with some complicated ideas and issues and, in order for him to be successful, he needed to find new ways to make his work accessible to more people. This is one of the primary motivations for his creation of the energy systems language he used to diagram the energy transformations in natural systems: it made it easier for everyone to see the big picture. This book attempts to do the same thing while providing arguments for new interpretations of scientific and philosophical concepts and positions.

This book also attempts to help us better understand how to handle a unique challenge that faces human societies around the world. Since the nineteenth century, industrial societies have been growing in an unprecedented fashion that has been fueled by a uniquely powerful and efficient form of energy: fossil fuels. This source of energy is, however, finite, and we are quickly approaching the peak of our ability to use fossil fuels. The use of this source of energy has also dramatically increased the carbon dioxide concentrations in our atmosphere and affected many other aspects of our environment causing global changes in the climate of the earth. We desperately need to find other sources of energy that do not adversely impact our environment and undermine the sustainability of human societies. This book illustrates how moral and religious values evolve in a manner that is fundamentally shaped by the thermodynamic principles that guide all natural systems. It can therefore help us better understand the changes that will be coming in the near future and how we can prepare for them.

Chapter 1 introduces the MPP and Nietzsche's concept of the will to power and briefly outlines some of the parallels between the work of Odum and Nietzsche. Chapter 2 provides a discussion of the work of the eighteenth- and nineteenth-century philosophers and scientists who developed early versions of the will to power or influenced Nietzsche's development of the concept. Chapter 3 describes Lotka's principle of maximum energy flux, his arguments for it, and how it revolutionized Darwin's theory of evolution, at least in certain circles (most biologists would not have heard of it). Chapter 4 explains the many different ways that Odum further developed Lotka's principle, renamed it the MPP, and how he used it to analyze all kinds of natural and physical systems, including human social systems. This chapter also discusses the empirical evidence we now have that supports the MPP. It also critically analyzes some of the criticisms of this principle. Chapter 5 describes Nietzsche's concept of the will to power, his argument for it, and it provides a textual argument for a unique interpretation of how he used the will to power in his critique of morality. This discussion of Nietzsche is placed after the discussion of Odum so that the philosophical implications of their critiques of moral and religious values can be considered after an empirical foundation for the discussion has already been established. Chapter 6 provides a detailed analysis of the parallels and differences between Nietzsche's use of the will to power and Odum's use of the MPP, and it considers the scientific and philosophical implications of their work and this investigation of it. This chapter ends with a discussion of the implications this investigation has for the issues associated with energy use: the depletion of fossil fuels, the role of renewable energy technology, and the impact a change in our ability to use energy can have on the evolution of moral and religious values.

Rockville, MD, USA   Timothy McWhirter

References

Brown, M. T. (2023). The maximum power principle. In E. P. Rosa & J. Ramos-Martin (Eds.), *Elgar encyclopedia of ecological economics* (pp. 363–367). Edward Elgar Publishing.

Glacier, D. S. (2024). Power and efficiency in living systems. *Science, 6*(28).

Hall, C. A. S. (Ed.). (1995). *Maximum power: The ideas and applications of H. T. Odum*. University Press of Colorado.

Hall, C.A.S. (2004). The continuing importance of maximum power. *Ecological Modeling, 178*, 107–113.

Lenton, T. M., Pichler, P., & Weiz, H. (2016). Revolutions in energy input and material cycling in Earth history and human history. *Earth System Dynamics, 7*(2), 353–370.

Lotka, A. (1922a). Contribution to the energetics of evolution. *Proceedings of the National Academy of Science, 8*(6), 147–151.

McWhirter, T. (2012). Nietzsche's naturalistic metaethics: In defense of privilege. *Philosophy Study, 2*(2), 92–102.

Nagel, T. (2012). *Mind and cosmos: Why the materialist neo-Darwinian conception of nature is almost certainly false*. Oxford University Press.

Odum, H. T. (1971). *Environment, power, and society*. Wiley.

Odum, H. T. (1982). Pulsing, power and hierarchy. In W. J. Mitsch, R. K. Ragade, R. W. Bosserman, & J. A. Dilon (Eds.), *Energetics and systems* (pp. 33–60). Ann Arbor Science Publishers.

Odum, H. T. (2007). *Environment, power, and society for the 21th century*. Columbia University Press.

Odum, H. T., & Odum, E. C. (1976). *Energy basis for man and nature*. McGraw-Hill.

Odum, H. T., & Odum, E. C. (2001). *A prosperous way down*. University Press of Colorado. Kindle version.

Odum, H. T., & Pinkerton, R. (1955). Time's speed regulator: The optimum efficiency for maximum power output in physical and biological systems. *American Scientist, 43*(2), 331–343.

Sciubba, E. (2011). What did Lotka really say? A critical reassessment of the "maximum power principle." *Ecological Modeling, 222*, 1348.

Steffen, W., Broadgate, W., Deutsch, L., Gaffney, O., & Ludwig, C. (2015). The trajectory of the Anthropocene: The great acceleration. *The Anthropocene Review, 2*(1), 81–98.

Acknowledgments

I would like to thank Charlie Hall for seeing the value of this project and providing me with the editorial guidance to take advantage of this opportunity. I also received valuable input on this manuscript from the systems ecologist Pat Kangas, the physicist Garvin Boyle, and the philosophers Christian J. Emden, Helmut Heit, Vanessa Lemm, and Claus Zittel. My research assistant, Ethan Mao, also provided valuable help. Any errors that remain are my fault.

Contents

1 Introduction 1
    1.1 The Maximum Power Principle 1
    1.2 Odum and Nietzsche and the Evolution of Power 6
    References 9

2 The Historical Roots of Maximum Power 11
    2.1 Introduction 11
    2.2 Maximum Power in the History of Science and Philosophy 12
          2.2.1 Maximum Power in the History of Ecological
                 Economics 15
    References 19

3 Lotka's Principle of Maximum Energy Flux 21
    3.1 Introduction 21
    3.2 Contributions to the Energetics of Evolution 22
    3.3 Natural Selection as a Physical Principle 25
    3.4 The Law of Evolution as a Maximal Principle 26
          3.4.1 Human Behavior and the Principle of Maximum
                 Energy Flux 31
    3.5 Critical Analysis 35
          3.5.1 Exosomatic Evolution and the PMEF 35
          3.5.2 Abiotic Systems and the PMEF 37
    3.6 Conclusion 38
    References 39

**4 Howard Odum and the Evolution of the Maximum Power
Principle** 41
    4.1 Introduction 41
    4.2 The MPP and Systems Ecology 42
    4.3 The Efficiency and Speed of Maximum Power 44
    4.4 Maximum Useful Power Output 47

        4.4.1   Power Output   47
        4.4.2   Useful Power   48
    4.5   Feedback Loops   48
    4.6   The Pulsing Paradigm   49
    4.7   Maximum Empower   50
    4.8   Definitions of the MPP.   52
    4.9   The Maximum Entropy Production Principle   53
    4.10 The Evidence   53
    4.11 Theoretical Support   58
    4.12 The Critics of the MPP   59
        4.12.1 The Problems with Empower.   59
        4.12.2 Reductive Interpretations of the MPP   60
        4.12.3 The MPP Vs. The MEPP   63
    4.13 The MPP and Morality, Religion and Politics   64
        4.13.1 Human Power.   64
        4.13.2 Moral and Religious Values   65
    4.14 Conclusion   68
    References   69

5   Nietzsche's Will to Power   73
    5.1   Introduction   73
    5.2   The Will to Power   76
        5.2.1   Nietzsche's Description of the Will to Power   76
        5.2.2   Nietzsche's Argument for the Will to Power   79
        5.2.3   Influential Interpretations of the Will to Power   81
    5.3   Nietzsche's Metaethical Epistemology   83
        5.3.1   The Will to Power as an Empirical Principle   85
        5.3.2   The Will to Power and Social Growth   85
        5.3.3   Nietzsche's "Science of Morals"   89
        5.3.4   Influential Interpretations of Nietzsche's Metaethics   93
        5.3.5   Moore's Open Question Argument   98
    5.4   Conclusion   98
    References   99

6   Odum and Nietzsche: Parallels, Differences and Implications   103
    6.1   Introduction   103
    6.2   Parallels   104
        6.2.1   General Systems Theory   104
        6.2.2   Nietzsche, Odum and Naturalism   107
        6.2.3   Change Is Essential to Nature   108
        6.2.4   The MPP and the OWP   109
    6.3   Differences   117
        6.3.1   War   118
        6.3.2   Democracy   120

6.3.3 The Evolution of Morality 121
6.3.4 The Metaethics of the Science of Morals 124
6.4 Implications 125
6.4.1 The MPP and the MEPP 125
6.4.2 MPP and the Minimum Entropy Production Principle 126
6.4.3 The Empirical Evidence for the MPP 126
6.4.4 The Evolution of the "Science of Morals". 127
6.4.5 Nietzsche and General Systems Theory 130
6.4.6 Nietzsche's Naturalistic Metaethical Epistemology 130
6.4.7 Reevaluating the Slave Rebellion in Morality 131
6.4.8 The Empirical Plausibility of the OWP 131
6.4.9 The Evolution of Knowledge. 132
6.4.10 The Evolution of Philosophy 133
6.4.11 Life After the Peak. 134
References. 138

7 Conclusion 143
References. 146

Chapter 1
Introduction

Abstract This chapter provides a brief overview of the book and a few reasons why it will be of interest to scientists, philosophers and readers with an interest in science and philosophy.

1.1 The Maximum Power Principle

In the nineteenth century, Darwin revolutionized our understanding of life, arguing that it evolves through a process of natural selection. His theory transformed biology and had a deep and abiding impact on science, philosophy, and cultures around the world. By comparison, most people are unaware of how Darwin's theory was further developed by scientists who viewed his work from the perspective of thermodynamics, in what has come to be called the "thermodynamic school" of evolution (Fry, 1995, 227): These scientists include Ludwig Boltzmann, Alfred Lotka, H. T. Odum, B. H. Weber, D. J. Depew, C. Kyke, S. N. Salthe, E. D. Schneider, and R. E. Ulanowics. In 1886 Boltzmann wrote,

> The general struggle for existence of animal beings is therefore not a struggle for raw materials—these, for organisms, are air, water and soil, all abundantly available—nor for energy, which exists in plenty in any body in the form of heat, but a struggle for (low) entropy, which becomes available through the transition of energy from the hot sun to the cold earth. (1886)

In 1922, the mathematician, physical chemist, and statistician Alfred Lotka argued that "the advantage must go to those organisms whose energy-capturing devices are most efficient in directing available energy into channels favorable to the preservation of the species" (1922a). As a result, natural selection tends to increase the amount of energy that cycles through organic systems. He referred to this tendency as the *principle of maximum energy flux* (PMEF), and argued that it applied to the selection of design features in all natural systems that develop out of equilibrium (Lotka, 1922a).

Later in the twentieth century, the prolific ecologist Howard T. Odum developed Lotka's PMEF further and used it to analyze all kinds of physical, biological,

© The Author(s), under exclusive license to Springer Nature
Switzerland AG 2025

T. McWhirter, *Maximum Power and its Philosophical Roots*, SpringerBriefs in
Energy, https://doi.org/10.1007/978-3-031-80622-3_1

1

ecological and economic systems. He eventually renamed this principle the *maximum power principle* (MPP) (1976, 1977, 1983, 2007). He worked initially with the physicist Richard C. Pinkerton to develop an argument that natural systems tend to operate at a level of efficiency that maximizes "power output" (Odum & Pinkerton, 1955). Odum developed an energy systems language that enabled him to describe the flows of energy through different systems and illustrate how they tend to maximize power. In a number of books and articles, Odum went on to apply the MPP and his unique form of energy analysis to all kinds of natural systems, including human social systems (1971, 1977, 2007; Odum & Odum, 2001).

Odum had a knack for taking complex concepts and making them readily understandable. For example, he defined the first and second laws of thermodynamics as follows: the first law says that the quantity of energy is conserved; the second law says "but the quality is not." Hence the second law tells us how we can do less total work over time as energy is transformed from one type to another.

Power measures the rate at which energy is expended in the form of work over time: $P = W/T$. When energy is expended, entropy is produced. So, increases in power are fundamentally associated with increases in the rate of the production of entropy. There are some processes that produce entropy without performing work, e.g., radiative transfer or diffusion, but all processes that perform work produce entropy. In the late twentieth century, scientists began to focus on how evolutionary processes maximize the production of entropy, and they began to use a new principle to refer to this tendency that they called the *maximum entropy production principle* (MEPP) (Paltridge, 1978). The MPP and the MEPP have had a unique impact on 20th and twenty-first century science. They have been used to describe phenomena in a number of very different scientific disciplines: ecology, earth system sciences, biology, chemistry, physics, sociology, and cosmology, among others. Scientists have suggested that these principles along with the principle of natural selection should be considered a fourth law of thermodynamics (Lotka, 1922b, p. 153), or a corollary of the second law (Martyushev & Seleznev, 2006, 3).

Surprisingly, the history of the development of the PMEF, the MPP and the MEPP have often been overlooked in contemporary scientific discussions. The Dutch-born paleoecologist and evolutionary biologist Geerat J. Vermeij at the University of California, Davis recently published a book with Princeton University Press entitled *The Evolution of Power: A New Understanding of the History of Life* that describes a version of the MPP without mentioning any of Odum's work on it (2023). This failure to take into account the work of scientists in the thermodynamic school of evolution has undermined the ability of contemporary scientists to accurately determine what is a "new" theory or principle.

The PMEF, the MPP, and the MEPP also have a heritage within the discipline of philosophy that has gone mostly unrecognized. The nineteenth century German philosopher Friedrich Nietzsche viewed his concept of the *will to power* as an empirical principle that describes how organic and inorganic systems develop in ways that increase their power, and he argued that we have good reason to consider whether the will to power can be used to describe all natural phenomena (Nietzsche, 1989, aphorism 36). Nietzsche was familiar with the work of scientists in the eighteenth and nineteenth centuries that were investigating different aspects of this thesis. Nietzsche used his concept of the will to power to guide his critique of philosophy, morality, science, religion, and culture.

Nietzsche is perhaps best known for his critique of morality. In this book, I develop a unique interpretation of his critique which holds that he did not evaluate the moral values from the perspective of some other set of moral values; he evaluated moral values from a *scientific* perspective: He described how moral values undermined or fostered the growth and flourishing of life as outlined by the will to power understood as an empirical principle. Odum critically examined moral and religious values almost a 100 years later in a similar fashion using the MPP (1971, 1977, 2007). The unique interpretation of Nietzsche's critique I develop brings into relief the striking parallels between the work of Nietzsche and Odum which have not been explored before. Our understanding of the nature and significance of these parallels will be enhanced by thoroughly investigating the differences that remain between Nietzsche and Odum. Exploring the relation between these two influential figures will be a primary focus of this book.

Exploring this relation will provide a number of benefits. First, it will help demonstrate the empirical plausibility of the will to power. The version of the will to power briefly described above has been viewed by philosophers as being empirically implausible. In Nietzsche's own time, the concept was largely ignored, along with much of his work. In the twentieth century, Nietzsche's work went on to have a profound impact on philosophy and culture. Jurgen Habermas describes him as the "turning point" that leads to "postmodernity" (Habermas, 1987, Ch. 4). Postmodern philosophers, like Michel Foucault, Jacque Derrida, and Gilles Deleuze, embraced Nietzsche's work with open arms. Nevertheless, Nietzsche's concept of the will to power was still viewed by philosophers as being empirically implausible. In retrospect, we can now see that philosophers were not aware of the work being done in the "thermodynamic school" of evolution studies. This view that the will to power was empirically implausible, oddly enough, came to be something which philosophers generally agreed on, even though they disagreed about many other aspects of Nietzsche's work. As postmodernism faded, Nietzsche began to be interpreted more and more, particularly among Anglophone philosophers, as a naturalist who used scientific methods to address philosophical questions. In spite of this naturalist turn in the secondary literature, philosophers continued to describe the version of the will to power outlined above as being incompatible with the "content of current science" (Clark, 2000, p. 120), and a "crackpot" view of metaphysics (Leiter, 2013, p. 594). This view has led leading figures in the secondary literature to make major decisions about the course of their research and, in some cases, it has led philosophers to develop "esoteric" interpretations of Nietzsche's work that claim to demonstrate that he never intended for the version of the will to power described above to be a part of his philosophy, notwithstanding the fact that he provided an argument for this concept in his published work—aphorism 36 of *Beyond Good and Evil*.

None of the philosophers who have questioned the empirical plausibility of the will to power have mentioned its relation to the scientific literature on the MPP or the MEPP. I was the first philosopher to mention the relation between the will to power and the MPP, just over ten years ago (McWhirter, 2012). This book will provide philosophers a completely different view of the empirical plausibility of the will to power. My assessment of Nietzsche's concept of the will to power provides evidence that the MPP may be *both* one of the most important *and* one of the least understood principles in contemporary science: It guides the evolution of all things and many well-educated people know nothing about it. This book seeks, among other things, to rectify this situation.

A second benefit of this book is that it can help both scientists and philosophers better understand and appreciate the philosophical implications and significance of Howard Odum's work. His energy systems analysis of morality raises a number of philosophical questions. For example, the philosopher G. E. Moore used logic and a form of conceptual analysis to argue that the supposition that 'goodness' can be defined in naturalistic terms, like pleasure or desire or power, involves a fallacy: the 'naturalistic fallacy.' Moore argued that if one claims that 'goodness' is defined in a particular way, like goodness is what maximizes power, we can tell that this claim is not true *by definition* because we can always doubt whether the claim is true: Its truth is always an 'open question.' By comparison, if one claims that $2 + 2 = 4$, the claim is true by definition and therefore we are not in a position to doubt it. Nietzsche's work provides some clues for how Odum's critique of morality can avoid Moore's 'open question' argument.

A third benefit of this investigation, which should be of particular interest to scientists, is that it corrects some contemporary misconceptions of the MPP. As mentioned in the preface, Odum sometimes described the optimum efficiency for maximizing useful power output increasing as the available energy decreases (Odum & Pinkerton, 1955; Odum & Odum, 2001, Odum, 2007). This point and its significance are overlooked by some contemporary scientists. For example, the professor of Energy Systems Enrico Sciubba concludes that the MPP is unable to accurately describe the development of natural systems both near and far from equilibrium (2011). He suggests that contemporary scientists tend to believe that two principles are needed to explain the development of natural systems: One principle is needed to describe how systems close to equilibrium develop in a manner that maximizes efficiency; another is needed to describe how systems far from equilibrium develop in a manner that maximizes the dissipation energy. This investigation demonstrates (in Chap. 4) that both of these principles are consistent with the MPP when it is interpreted accurately (Odum, 1982).

A fourth benefit of this investigation is that it can bring into focus certain problems that face the practice of philosophy now and in the future. Many contemporary philosophers consider themselves to be naturalists of one form or another. The term is hotly debated, but one common definition of philosophical naturalism is that it refers to a tendency to use scientific methods to address philosophical questions. The evidence suggests that the response of philosophers to Nietzsche's concept of the will to power is much more than merely an oversight: It illustrates that philosophers have not had the ability to accurately determine what is or is not empirically plausible according to the sciences in the nineteenth, twentieth, or twenty-first centuries. This is a problem.

There are certain aspects of this problem that may be specific to Nietzsche scholarship. Nietzsche did not faithfully cite his sources or develop detailed arguments in a systematic fashion. He made extravagant claims, told jokes, and claimed to have unique insight. He usually wrote in an aphoristic style where he jumped from point to point. This may have made it easier for philosophers and others to dismiss his claims about the version of the will to power outlined above. Some philosophers in the secondary literature characterize Nietzsche's unique writing style as a strength of his work, which keeps it from settling into a systematic, modern form of metaphysics; however, one could argue that his style of writing was the Achilles heel of a great philosopher, which prevented his readers from clearly seeing, understanding, and appreciating the extraordinarily prescient idea that guided much of his work—the will to power.

There are some aspects of this problem that may be a unique product of the research on the MPP and the MEPP. Odum began to develop the MPP further with Pinkerton in their 1955 paper, but they generally refer to it there as "*maximum power output*" (1955). In his 1971 edition of *Environment, Power, and Society*, Odum refers to "*maximum power*" twice (1971, pp. 30, 163), and he makes no explicit reference to the *maximum power principle*. The first time he uses the name—the *maximum power principle*—is in his 1976 book *Energy Basis for Man and Nature* (Odum & Odum, 1976). He continues to use this name from this point on (1983), but he also began to refer to what he calls the *maximum empower principle*, which is a further development of the MPP that we will consider in Chap. 4, Sect. 4.7 (1995). In his 2007 edition of the book, *Environment, Power, and Society in the Twenty-First Century*, he refers to the "*maximum power principle*" six times (2007, 32, 56, 59, 89, 91, 128), "*maximum power*" 44 times, "*maximum empower principle*" 5 times, and "*maximum empower*" 16 times.

In 1978, the scientist G. W. Paltridge used the MEPP to describe the evolution of atmospheric systems (1978). Almost a decade later, other scientists suggested that ecosystems evolved in a manner that is guided by the MEPP, but their work did not receive a lot of attention (Ulanowicz & Hannon, 1987; Swenson, 1989). In 2003 and 2005, the broadly trained scientist Roderick Dewar provided a theoretical basis for the MEPP and derived a provisional proof of it for non-equilibrium, steady-state systems, both biotic and abiotic (2003, 2005). In 2006, the scientists L. M. Martyushev and V. D. Seleznev described how the MEPP has been used to describe phenomena in several different scientific disciplines, including physics, chemistry, and biology (2006). So, by the time Odum began to actually refer to the "*maximum power principle*," many scientists had already left it behind in favor of the MEPP. Odum also eventually began to leave it behind himself, to a certain extent, in order to use the *maximum empower principle*. While the possible relation between the will to power and the MPP would be obvious to a philosopher, the relation between the will to power and the maximum empower principle would not be so clear, and the relation between the will to power and the MEPP would be even less clear. This could, to some extent, explain why contemporary philosophers failed to see the empirical plausibility of Nietzsche's concept of the will to power.

Some aspects of this problem may be a product of the growth of science itself, and this can present a challenge to the prospects for philosophical naturalism. Since science split off from philosophy as a separate discipline, the number of scientific disciplines and the amount of work being done in them has increased exponentially. There may have been a time in the eighteenth century when a renaissance scholar could take within her compass all the latest developments in the sciences, but those days are long gone. Consequently, the challenge for a discipline that seeks to be the science of sciences—like philosophical naturalism—has increased in an exponential fashion. The training that philosophers receive in graduate school may need to adapt in order for them to be better prepared to meet this challenge. This book will not provide a solution to this difficult problem, but it will provide a clear and important illustration of it.

A fifth benefit of this book is that it will begin to explain the relation between the evolution of moral and religious values and our use of energy, at a unique time when our use of energy is in the process of undergoing dramatic changes. As mentioned briefly in the preface, since the nineteenth century, industrial societies have been able to grow in an unprecedented fashion because they have been able to use an extremely efficient and powerful source of energy: fossil fuels. Many scientists believe we are quickly approaching the peak of our ability to access and use fossil fuels. On the other side of this peak, things are going to change in ways that could impact the ability of societies and economies to function and grow, their ability to maintain order, and the evolution of moral and religious values. Nietzsche and Odum provide a unique view of the relation between the evolution of moral and religious values and power that can help us negotiate our way through the challenges in the offing.

1.2   Odum and Nietzsche and the Evolution of Power

This book will highlight a comparison of Odum's use of the MPP and Nietzsche's use of the will to power to describe the evolutionary development of various systems, including ecosystems and human societies. It will illustrate how they bring power and energy into a discussion of disciplines well beyond physics. Most of human thought evolved and was written when humans had no real idea about how energy worked beyond some obvious things: plants needed sunlight, muscular people and animals could do more work than others etc. In the middle of the nineteenth century, things began to change with the work of the engineer Sadi Carnot, the physicist Julius Robert von Mayer (1814–1878), and the physicist, mathematician, and brewer James Prescott Joule (1818–1889). They began to develop a better understanding of how energy worked. During this same period, Darwin also transformed the sciences with his view of evolutionary theory. So, the nineteenth century brought these two intellectual revolutions: thermodynamics and evolution theory.

Yet our understanding of moral and religious values was still based on the writings of philosophers and theologians that lived long before these revolutions. Nietzsche and Odum begin to develop a "science of morality" that takes these two revolutions into account.

This book will demonstrate that while Odum and Nietzsche disagree on many important issues, they agree on metaethical and metaphilosophical issues: They agree on the big picture. As mentioned, I will maintain that Nietzsche argues that we should analyze moral and religious values from a scientific perspective based on his concept of the will to power. Odum argues that we should analyze moral and religious values from a scientific perspective based on the MPP. Nietzsche and Odum generally agree on the perspective we should use to analyze moral values and the principle that guides evolutionary development.

Odum and Nietzsche also share a similar view of the concept of power [*Macht*]. They both see this concept applying to all natural systems, both biotic and abiotic, and human social systems. In English, the same word is used to describe the scientific concept of power used to analyze natural systems, as well as the concept of power used to describe the economic, political, and military power manifested by human social systems, but the two concepts of power are usually considered to be distinct. Odum seeks to emphasize the similarities between these concepts: He argues that the different forms of power manifested by human social systems are structured by thermodynamic laws and they can be measured just like the power of other natural systems. In German, there are two different words for the English word "power": "*Leistung*", which is the scientific concept of power (power = work/time); and "*Macht*", which refers to the ability of humans to influence other humans, e.g., political power. Nietzsche describes his concept of the will to power using the term "*Macht*": "*der Wille zur Macht*". But he describes the will to power applying to all natural systems, both biotic and abiotic, human social systems, individual human beings, and the entire "world". Consequently, Odum and Nietzsche both use the concept of power [*Macht*] in a way that undermines or deconstructs the distinction between "*Leistung*" and "*Macht*".

When we consider the perspective that both Odum and Nietzsche used to analyze moral and religious values in more detail, it helps us better understand the many similarities and differences between them. Since the perspective they use is scientific, it will evolve over time like other scientific disciplines. Nietzsche, in a real sense, was like a nineteenth century social physician that, like Freud, tried to create a whole new science of social health. He had a vague idea of what he was trying to do; his positions were based largely on intuition and prejudices; there was hardly any research he could use to guide his work; he had to use a vanity publisher to publish most of his books; he was not really trained to practice this science; he was a philologist who became a philosopher who reads everything. We might say he was trained to see the potential of this science rather than practice it.

Odum, on the other hand, was a twentieth century systems ecologist, and some might say social physician, that developed cutting-edge tools and a whole new language for this science of social health; he was eminently trained to do this work; in many cases, he based his positions on a wealth of research that he developed with colleagues and graduate students; he published his work in journals all around the world; many of his papers helped start new scientific disciplines. The seventeenth century physician William Griggs diagnosed young "afflicted" girls as witches under the influence of the devil. Physicians in subsequent centuries diagnosed illnesses in different ways and prescribed different therapies. The science of medicine evolves over time and we now expect doctors to treat illnesses in ways that are much more effective than just a few decades ago. This book will argue that Nietzsche's specific positions on moral and political issues should be viewed like the prescriptions of a nineteenth century social physician; Odum's views are those of a twentieth century social physician. They are both practitioners of a new "science of morality" that views moral and religious values as products of an evolutionary process that is guided by the will to power, or what we could also call the maximum power principle.

When we compare the unique work of these two revolutionary thinkers, it can bring into relief inconsistencies in their respective positions that we may not have noticed otherwise. This can enable us to develop new ways to overcome these inconsistencies and bring to life a new view that is implicit in their work. For example, Nietzsche describes evolution being guided by the will to power, but he also describes the evolution of morality undermining the will to power: he argues that the emergence of what he calls the *"slave revolt in morality"* undermined the growth of power (*GM* I 7). When you compare his view of the evolution of morality with Odum's, this inconsistency becomes more obvious, as well as the way to overcome it.

Odum describes moral and religious values evolving over time to meet the different energy demands of societies at different times, but he does not mention in any real detail the processes of cultural evolution that are involved in this evolution. The scientific study of cultural evolution has progressed to a point now where we can add more detail to this view. When we do, it can add more weight to a related question: How does scientific knowledge evolve? Odum does not say much about this, but his work and the work of scientists studying cultural evolution can help us address this question.

This book will seek to synthesize the ideas of Nietzsche and Odum. In the process, it will develop a view that, to some degree, goes beyond what they have written. One of the more surprising things I found in my research for this book is the extent to which Nietzsche's work is consistent with the principles of general systems theory. The relation between Odum's work and general systems theory is well known; this is not the case with Nietzsche's work. This parallel will be discussed in greater detail in the book. One fundamental tenet of systems theory is that 'the whole is greater than the sum of its parts.' This applies to the synthesis of Nietzsche's and Odum's ideas outlined in this book.

This synthesis has a particular relevance now. One of the arguments for general systems theory in this information age is that the sciences are generating unprecedented amounts of data about specific topics in specialized disciplines, but we do not necessarily have an effective way to understand the relation between all this data: We know a great deal about the leaves, but we cannot clearly see the forest. For a number of reasons, it is getting more important for us to see the forest. First of all, many scientists are now recognizing that human activity has had a dominant influence on the planet. This is why many scientists now refer to the present geological age as the *Anthropocene* (Steffen et al., 2015). Second, scientists have also recognized that human activity is causing global climate change. We see the effects every day: record temperatures, more frequent and destructive storms, rising sea levels, and various threats to agricultural production. Third, industrial societies have grown in a nonlinear fashion since the nineteenth century, in large part, because of their ability to use fossil fuels, and we are quickly approaching the peak of our ability to use this finite energy resource. The synthesis between thermodynamics, ecology, evolutionary theory, and morality provided by Nietzsche and Odum and their concepts of the will to power and the MPP enable us to understand the big picture more clearly at this particularly crucial time.

Chapter 2 will begin this investigation of the roots of the MPP by describing the work of the eighteenth and nineteenth century philosophers and scientists who developed early versions of the will to power or influenced Nietzsche's development of the concept.

References

Boltzmann, L. (1886). The second law of thermodynamics. In B. McGinness (Ed.), *Ludwig Boltzmann: Theoretical physics and philosophical problems: Selected writings* (pp. 14–32). D. Reidel.

Clark, M. (2000). Nietzsche's doctrine of the will to power: Neither ontological nor biological. *International Studies in Philosophy, 32*(3), 119–134.

Dewar, R. (2003). Information theory explanation of the fluctuation theorem, maximum entropy production and self organized criticality in non-equilibrium stationary states. *Journal of Physics A: Math and General, 36*, 631–641. https://doi.org/10.1088/0305-4470/36/3/303

Dewar, R. (2005). Maximum entropy production and the fluctuation theorem. *Journal of Physics A: Math and General, 38*, L371–L381. https://doi.org/10.1088/0305-4470/38/21/L01

Fry, I. (1995). Evolution in thermodynamic perspective: A historical and philosophical angle. *Zygon, 30*(2), 227–248.

Habermas, J. (1987). *Der philosophische Diskurs der Moderne* [The philosophical discourse of modernity]. (F. Lawrence, Trans.). Polity Press. Kindle Edition.

Lieter, B. 2013. Nietzsche's naturalism reconsidered, in Ken Gemes / John Richardson (eds.) The Oxford handbook of Nietzsche. Oxford University Press. 578–598.

Lotka, A. (1922a). Contribution to the energetics of evolution. *Proceedings of the National Academy of Science, 8*(6), 147–151.

Lotka, A. (1922b). Natural selection as a physical principle. *Proceedings of the National Academy of Science, 8*(6), 151–154.

Martyushev, L. M., & Seleznev, V. D. (2006). Maximum entropy production principle in physics, chemistry and biology. *Physics Reports, 406*(1), 1–45.

McWhirter, T. (2012). Nietzsche's naturalistic metaethics: In defense of privilege. *Philosophy Study, 2*(2), 92–102.

Nietzsche, F. (1989/1886). *Beyond good and evil* [Jenseits von Gut und Böse. Vorspiel einer Philosophie der Zukunft]. (W. Kaufmann, Trans.). Vintage.

Odum, H. T. (1971). *Environment, power, and society.* Wiley.

Odum, H. T. (1977). The ecosystem, energy, and human values. *Zygon, 12*(2), 109–133.

Odum, H. T. (1982). Pulsing, power and hierarchy. In W. J. Mitsch, R. K. Ragade, R. W. Bosserman, & J. A. Dilon (Eds.), *Energetics and systems* (pp. 33–60). Ann Arbor Science Publishers.

Odum, H. T. (1983). *Systems ecology: An introduction*. Wiley.

Odum, H. T. (1995). Self-organization and maximum empower. In C. A. S. Hall (Ed.), *Maximum power: The ideas and applications of H. T. Odum*. University Press of Colorado.

Odum, H. T. (2007). *Environment, power, and society for the 21th century*. Columbia University Press.

Odum, H. T., & Odum, E. C. (1976). *Energy basis for man and nature*. McGraw-Hill.

Odum, H. T., & Odum, E. C. (2001). *A prosperous way down*. University Press of Colorado. Kindle version.

Odum, H. T., & Pinkerton, R. (1955). Time's speed regulator: The optimum efficiency for maximum power output in physical and biological systems. *American Scientist, 43*(2), 331–343.

Paltridge, G. W. (1978). The steady-state format of global climate. *Quarterly Journal of the Royal Meteorological Society, 104*, 927–945. https://doi.org/10.1002/qj.49710444206

Sciubba, E. (2011). What did Lotka really say? A critical reassessment of the "maximum power principle". *Ecological Modeling, 222*, 1348–1353.

Steffen, W., Broadgate, W., Deutsch, L., Gaffney, O., & Ludwig, C. (2015). The trajectory of the anthropocene: The great acceleration. *The Anthropocene Review, 2*(1), 81–98.

Swenson, R. (1989). Emergent attractors and the law of maximum entropy production: Foundations to a theory of general evolution. *Systems Research, 6*, 187–197.

Ulanowicz, R. E., & Hannon, M. (1987). Life and the production of entropy. *Proclamations of the Royal Society of London, 232*, 181–192. https://doi.org/10.1098/rspb.1987.0067

Vermeij, G. J. (2023). *The evolution of power: A new understanding of the history of life*. Princeton University Press.

Chapter 2
The Historical Roots of Maximum Power

Abstract This chapter traces the early roots of the maximum power principle in eighteenth and nineteenth century science and philosophy. It demonstrates that many scientists (and others) saw something like the MPP in the systems they studied, and that the idea was "in the air."

2.1 Introduction

H. T. Odum describes the maximum power principle (MPP) as follows, "systems that prevail are those with loading adjusted to operate at the peak of the power efficiency curve .... During self-organization, these systems reinforce (choose [through natural selection]) pathways with the optimum load for maximum output" (2007, p. 37). He believed that the evolution of organic and inorganic systems was guided by this principle. Whenever Odum describes this principle, he generally cites Alfred Lotka as the scientist who first proposed it (1955). In the short essay in which Lotka first proposed the principle (1922), which he referred to as the *principle of maximum energy flux*, he begins by pointing out that Ludwig Boltzmann suggested that the fundamental object of contention in the evolution of life is available energy (1886), and in a footnote Lotka also cites the work of Walther Nernst (1913), David Burns and D. Noel Patton (1921), and Henry Osborn (1918). Lotka does not, however, suggest that any of these scientists proposed that natural selection is guided by the MPP or the *principle of maximum energy flux*.

Friedrich Nietzsche describes the will to power guiding the evolution of all organic and inorganic systems in a manner that increases their power (1886, aphorisms 13, 36). While he develops this concept within the context of his critical analysis of other concepts, such as Schopenhauer's concept of the will to life, Nietzsche does not describe in his published writings any other philosopher or scientist proposing a concept that is the same or similar to the will to power. One could be led to believe that Lotka was the first scientist to propose a version of the MPP and that Nietzsche was the first philosopher to propose the will to power.

© The Author(s), under exclusive license to Springer Nature Switzerland AG 2025

T. McWhirter, *Maximum Power and its Philosophical Roots*, SpringerBriefs in Energy, https://doi.org/10.1007/978-3-031-80622-3_2

However, in an essay that Odum wrote late in his career (1995, p. 311), he mentions that the economist Joan Martinez-Alier provides evidence that ideas similar to the MPP were proposed by scientists in the nineteenth century (1987). In his effort to describe the history of ecological economics, Martinez-Alier uncovers a number of ideas and views related to the MPP in the work of Eduard Sacher, Patrick Geddes, Leopold Pfaunder, Henry Adams, and Wilhelm Ostwald.

In his discussion of the will to power, Nietzsche endorses the eighteenth century physicist Roger Joseph Boscovich's (1711–1787) view of matter as fields of force, but Nietzsche does not cite the work of the many other scientists and philosophers who he had clearly read and who influenced his work on the will to power. Many philosophers have dismissed Nietzsche's concept of the will to power as being empirically implausible without considering in any detail the work of those scientists in Nietzsche's time that supported it. There are some notable exceptions here. Christian Emden provides an extremely detailed investigation of the relation between Nietzsche's philosophy and the work of scientists in the Nineteenth Century Life Sciences (2014). Gregory Moore describes how deeply Nietzsche was immersed in the debates scientists were having about evolution in the late nineteenth century (2002). From their work, we learn that the French philosopher Jean-Marie Guyau and the entomologist William Henry Rolph both proposed a view of life that is consistent with Nietzsche's concept of the will to power, and Maximilian Drossbach describes a concept of force that Nietzsche believed was similar to his own. Beyond all this, we can also find some concepts used by German philosophers in the eighteenth century that have some similarities to Nietzsche's concept of the will to power.

This chapter will focus on illustrating the roots of the MPP in the history of science and philosophy in the work of those other than Nietzsche, Lotka, and Odum. As we shall see, the roots of this principle run quite deep.

2.2   Maximum Power in the History of Science and Philosophy

We first begin to see concepts that relate to the MPP and the will to power in philosophy in 1790 when Immanuel Kant (1724–1804) published his *Critique of Judgment*, in which he described organic systems as "self-organizing beings" (Kant, 1951). By this, he meant their origin was not caused externally; it was the product of an internal capacity to produce organization. Kant was influenced by the physiologist and naturalist Johann Friedrich Blumenbach (1752–1840) and his concept of *Bildungstrieb* (formative drive) (1781; Gigantes, 2009). Friedrich Schelling (1775–1854) extended Kant's thesis to inorganic systems and argued there is a principle of organization in all organic and inorganic matter, and with this view he believed he offered an alternative to a mechanistic or vitalistic conception of life (Heuser-Kebler, 1986). In Chap. 5, Sects. 5.1 and 5.2.1, we will discuss how scientists eventually realized in the twentieth century that, in order for a natural system to develop its organization in a manner that is consistent with the second law of thermodynamics, it must develop its ability to extract, use, and then dissipate energy from its environment, i.e., it must develop its power: so we now have the ability to see the connection between the increase in organization and power. Kant and Schelling are two of the most famous representatives of the philosophical movement referred to as German Idealism. Some version of the will to power can be viewed as emerging naturally from this philosophical movement.

There were other philosophers and scientists whose work Nietzsche read that had a direct impact on his development of the will to power. The German philosopher Friedrich Lange (1828–1875) posited a "law of development" as a purposeful force that supplies possible forms from which natural selection chooses the actual forms (Lange, 1950 [1865]). George J. Stack's book *Lange and Nietzsche* (1983) describes Lange's considerable influence on Nietzsche. In the mid 1880s, Nietzsche read the philosopher Jean-Marie Guyau's (1854–1888) book *A Sketch of Morality Independent of Obligation or Sanction [Esquisse d'une morale sans obligation ni sanction]* (1885). Guyau argues that the central goal of human agency is the intensification of the experience of living (1885; also see Ansell-Pearson, 2009). He describes the "power [*puissance*] of life" as a productive force that assimilates resources from its environment and continually *expands* (1885). A similar view of life expanding was developed by the entomologist William Henry Rolph (1847–1883). Nietzsche read the book Rolph published in 1882 entitled *Biological Problems [Biologische Probleme]* (1884). Like Nietzsche, Rolph criticizes Herbert Spencer's interpretation of the application of natural selection to morality and social development; rather than Darwin's "struggle for existence" or Spencer's "survival of the fittests," Rolph argued that evolution involved a "struggle for the expansion of life [*Kampf um Lebensvermehrung*]" (Rolph, 1884). We can see Rolph's and Guyau's view that the power of life continually expands reflected in Nietzsche's statement: "physiologists should think before putting down the instinct of self-preservation as the cardinal instinct of an organic being. A living thing seeks above all to discharge its strength – life itself is *will to power*-: [...]" (1886, aphorism 13).

Nietzsche had a mixed view of the work of Julius Robert Mayer (1814–1878), a physician who made significant contributions to thermodynamics. Mayer published his book *The Mechanics of Energy [Die Mechanik der Wärme]* in 1867. Nietzsche was critical of Mayer's materialism, but he did agree with Mayer's theory of heat which provided support for Nietzsche's reading of Rolph. Nietzsche agreed with Rolph's view that the evolution of life was based on a "struggle for the expansion of life" which often wasted energy and used more of it than what is necessary to merely preserve life. Mayer investigated the energy wasted by cannons, steam engines, and animals and wrote, "The chemical process is always larger than its useful effect" (1845), which Nietzsche quoted with approval in his notes (Nietzsche, 1967, vol. 9, p. 451).[1] We will explore the relation between efficiency and the maximum power principle in more detail in Chap. 6, Sect. 6.2.4.

---

[1] Emden makes this point (2014, p. 173).

Nietzsche read Maximilian Drossbach's (1810–1884) book *On the Apparent and Real Causes of Events* [*Ueber die scheinbaren und wirklichen Ursachen des Geschehens*] shortly after its publication in 1884. There, Nietzsche found a concept of power, or force, that was similar to the one he was developing (See, for example, Schmidt, 1988). Drossbach writes that a "proper" understanding of force recognizes it as the "striving for expansion [*Streben nach Entfaltung*]" (1884). Nietzsche underlined these three words and wrote in the margins of his copy of Drossbach's book: "'will to power,' is what I say."[2]

Another important element of Drossbach's view, from Nietzsche's perspective, was that he thought the development of natural systems was driven by their interactions with other systems; it was not the product of some kind of inner drive. Drossbach writes:

> Natural beings [*Wesen*] develop their power by acting upon others and by meeting the agency of others. Reciprocal agency [*Wechselwirkung*] is the means for effective expansion, and the more complete the form of reciprocal agency, the more fully they [*i.e.*, natural beings] develop. (1884, 45)

Nietzsche knew that Roger Boscovich described matter as a product of an interaction between fields of force (1886, aphorism 12); he could see that Drossbach described power as a product of an interaction between natural beings; he went on to describe the mind as a product of an interaction between different drives. This parallel plays a major role in the most famous argument Nietzsche offers for the will to power (1886, aphorism 36).[3]

The contemporary philosopher Greg Moore argues that the will to power was based on a number of competing "non-Darwinian theories" (2002): among them, he includes the Swiss botanist Carl Nageli's (1817–1891) perfection principle, which held that evolution is directed by an "inner" tendency of organisms to perfect themselves; the German biologist Wilhelm Roux's (1850–1924) view that a struggle for existence must take place on the cellular or molecular level; and the German entomologist William Henry Rolph's view, which we have already discussed. Moore argues that although Nietzsche is critical of many aspects of nineteenth-century science and philosophy, in his concept of the will to power, Nietzsche still adopts a teleological view that reflects a problematic aspect of nineteenth-century science.

Moore's point is well taken, so far as it goes. We will discuss later how many of Nietzsche's views on different issues are influenced by the biases and prejudices of his time. That being said, Moore does not mention in his book any of the 20th and twenty-first century scientists in the "thermodynamic school" of evolution that describe natural selection being guided the principle of maximum energy flux, the MPP, or the maximum entropy production principle, which all entail some form of

---

[2] See the comment in Nietzsche's copy: Herzogin Anna Amalia Bibliothek, Weimar, Germany, Sig. C 252.

[3] The significance of the interactions between organisms in evolutionary processes is further developed by contemporary scientists. The zoologist Dolph Schulter, for example, has focused on demonstrating how "competitive interactions" seem to be a "common explanation for the diversity of forms." See his *The Ecology of Adaptive Radiation* (2000). Also see Hunt (2023).

a thermodynamic teleology: Lotka, Odum, Rod Swenson, Roderick Dewar, or Axel Kleidon. Moore does not explain how evolution can be consistent with the second law of thermodynamics and not entail some kind of thermodynamic teleology. He does not mention the second law of thermodynamics or the science of thermodynamics. Consequently, Moore is correct that Nietzsche's view of evolution entails a teleological component that is, in some sense, similar to other teleological views in the nineteenth century, but he is not in a position to judge whether this undermines or enhances the empirical plausibility of the will to power according to the contemporary sciences.

2.2.1   Maximum Power in the History of Ecological Economics

Some contemporary scientists believe that ecological economics emerged as a discipline in the 1990s. The economist Joan Martinez-Alier provides evidence that a number of scientists were doing work in this general area in the nineteenth century and many of them developed ideas that are similar to the MPP (1987). The MPP applies to all natural systems, both biotic and abiotic, so this focus on ideas developed in ecological economics is unique. The main reason we are focusing on these ideas here is because when Odum used the MPP, he usually cited Lotka as the source of it; Odum's reference to Martinez-Alier's work is the only time I am aware of that he refers to scientists working with ideas similar to the MPP before Lotka (1995, p. 311). So, this review of ideas related to the MPP in nineteenth century ecological economics can help us better understand Odum's view of the historical roots of the MPP.

Eduard Sacher (1834–1903) was a physics and mathematics teacher who wrote a book in 1881 entitled *Outline of a Mechanics of Society* [*Grundzüge einer Mechanik der Gesellschaft*], in which he writes, "The economic task of the available labour force consists of winning from nature the greatest possible amount of energy" (1881, p. 24; Martinez-Alier, 1987, p. 65). He measures the economic growth of societies in terms of their use of available energy. He provided the following statistics on the kilocalories available per person/per year for different groups in different historical periods (1881, p. 33; Martinez-Alier, p. 68):

'Savages'	three million
'Nomads'	six million
'Agriculturalists'	14 million
'Contemporary Central Europeans'	20.3 million

Martinez-Alier notes that Sacher does not provide much explanation for many of these figures, and there are some questions about their accuracy. However, for the purposes of this investigation, we can see the similarity between Sacher's description and the MPP: he describes social systems evolving in a way that increases their power.

The physicist Leopold Pfaundler (1839–1920) was one of the first to publish a realistic study of the carrying capacity of the earth (1902). The article was entitled, 'The World Economy in the light of Physics.' He argues that in order to understand the world economy, one needs to understand the 'law of entropy.' The total amount of energy does not change, as described by the first law of thermodynamics, but the 'quality' of energy changes over time, as described by the 'law of entropy.' Once heat energy is dissipated, it cannot do work. When an electric current passes through a conductor, Joule described how the current will be transformed into heat; however, this heat cannot entirely be transformed back into an electric current: much of the heat will be dissipated. Pfaundler recognized a distinction in the quality of energy: Helmholtz referred to energy that is capable of doing work as 'free energy;' Pfaundler referred to the dissipated energy that cannot do work as 'entropy' (Martinez-Alier, 1987, p. 114). Martinez-Alier notes that Pfaundler did not quote Classius, who had introduced the term "entropy." Pfaundler therefore revised Saucher's statement that the "economic task of the available labour force consists of winning from nature the greatest possible amount of energy" (1881, p. 24; Martinez-Alier, 1987, p. 65); Pfaundler argued that the economic struggle was for *free* energy.

Patrick Geddes (1854–1932) was a biologist initially and later became an urban designer. Along with Sacher, he was one of the first authors to correlate periods in human history with the rate at which energy was expended by social systems. He proposed the comparison of historical periods outlined below (1881; Martinez-Alier, 1986, p. 94) (Fig. 2.1). Martinez-Alier notes that Geddes did not describe this evolution in terms of "exponential curves but logistic ones" (1986, p. 94). Once again, we see a scientist describing social systems evolving in a way that increases their productive power.

Henry Adams (1838–1918) was a historian and a member of the Adams family, which produced two presidents of the United States, John Adams and John Quincy Adams. He appears to be the first professional historian to correlate historical periods in terms of the use of energy, and his work led him to outline a new law of human progress, which he called the "*law of acceleration.*" His articulation of this law in his autobiography is unique (1904a). The contemporary philosopher Eric Steinhart writes that Adams' original text, written in an exaggerated version of the language and style of his time, is "almost completely unreadable in 2011" (1904b). Steinhart provides an edited and paraphrased version for contemporary readers. With his law of acceleration, Adams argues that the progress of human societies represents a form of accelerating movement and, as outlined in Newton's first law of motion, this acceleration will continue unless it is acted upon by some other force. To illustrate this acceleration, Adams describes the growth of the coal-output in the nineteenth century: "The coal-output of the world, speaking roughly, doubled every ten years between 1840 and 1900, in the form of utilized power, for the ton of coal yielded three or four times as much power in 1900 as in 1840" (1904a; Martinez-Alier, 1987, p. 118). He suggests "the ratio of increase in the volume of coal-power may serve as dynamometer" (1904b, p. 2). In his life, he had "seen the coal-output of the United States grow from nothing to three hundred million tons or more" (1904b, p. 4). His law of acceleration describes this growth accelerating.

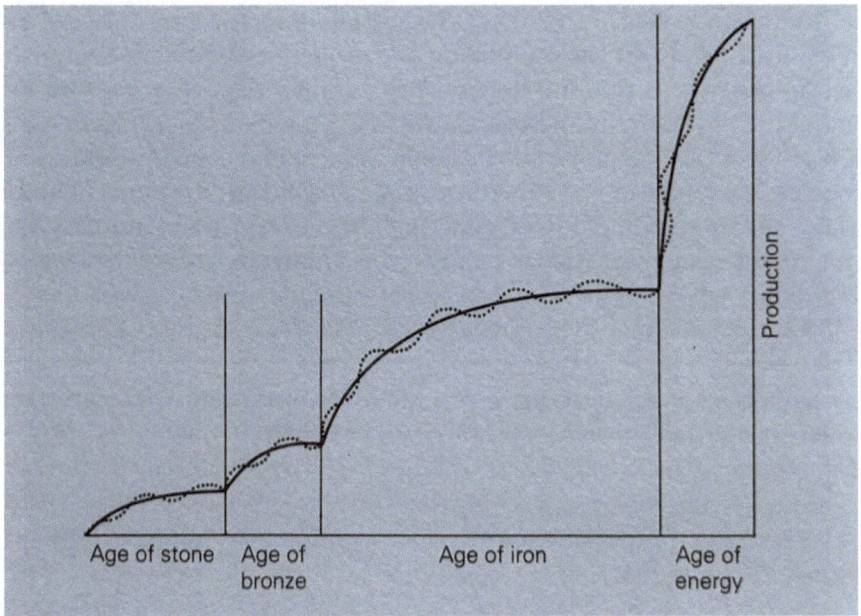

Fig. 2.1 The Comparison of Historical Periods provided by Patrick Geddes (1881). The age of stone is followed by the age of bronze, iron, and energy

Adams develops his view further in "A Rule of Phase Applied to History," in *Degradation of the Democratic Dogma* (1919). There, he argues that the law of acceleration leads humanity through a process of development that increases the power of production and entails four phases that are each characterized by a different relation between technology and humanity—the religious, mechanical, electrical, and ethereal. Each phase becomes increasingly shorter: a 90,000 year religious phase; 300 year mechanical phase; 17 year electrical phase; 4 year ethereal phase. Adams speculated that a simple "law of squares," much like Newton's inverse square law, could determine the length of each new phase. Unlike Geddes, Adams describes this evolution in terms of exponential curves rather than logistic ones (Fig. 2.2).

One can and should question the accuracy of Adams' speculations. He did not devote a great deal of time to compiling empirical evidence to support his views or even to clearly articulating them. Stanley Jevons had given empirical data and projections of exponential growth (and its limits) for coal in England in 1865 in "The coal question." However, Adams' general idea that the phases of development will have decreasing lengths appears to be prescient. Consider, for example, the evolution of the modes of production in human history: we have evidence that humans existed as hunters and gatherers for approximately 1.8 million years, at least; we used agriculture and the domestication of animals for 10,000–12,000 years; the first industrial revolution lasted approximately 110 years (it began in 1760 and the second industrial revolution began 1870); the second lasted approximately 77 years

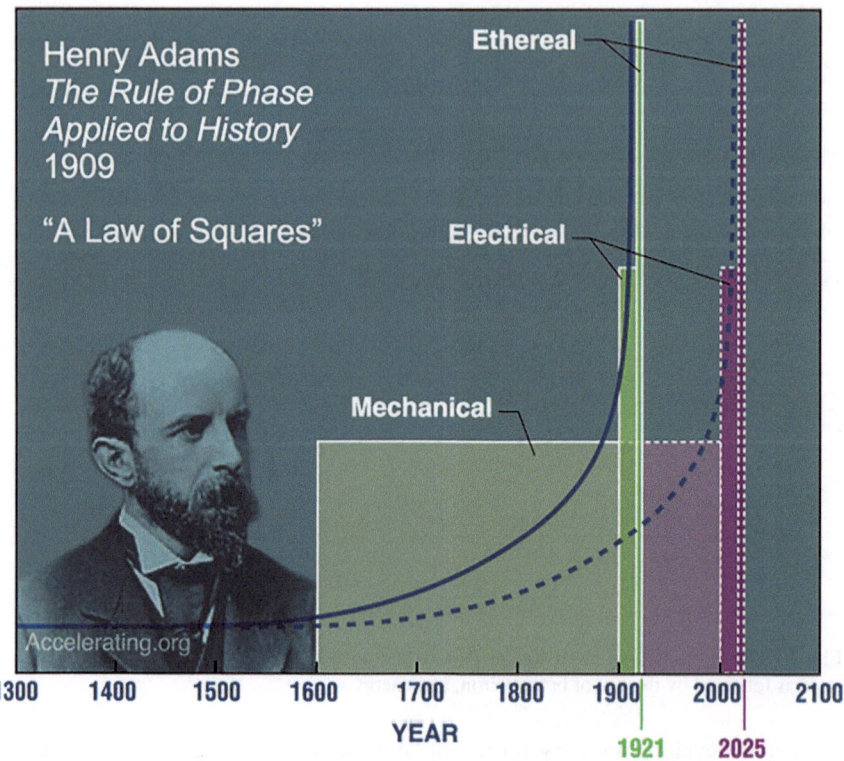

Fig. 2.2 The Rule of Phases Applied to History, by Henry Adams. The mechanical phase is from 1600 to 1900; the electric phase is from 1900 to 1917; the ethereal phase is from 1917 to 1921. The religious phase, which is 90,000 years long, is not shown completely

(the third industrial revolution began in 1947). The decreasing length of these phases provides compelling evidence that the rate of change is accelerating. In Chap. 4, Sect. 4.10, we will consider evidence compiled by contemporary scientists that supports Adams' "law of acceleration" (Steffen et al., 2015).

Wilhelm Ostwald (1853–1932) was a German chemist and philosopher who is credited with being one of the founders of the field of physical chemistry. He won a Nobel Prize in 1909 for his contributions to the fields of catalysis, chemical equilibria, and reaction velocities. He had a continuing interest throughout his career in the unification of the sciences through systematization, with a particular focus on what he considered to be the fundamental role played by energy. In 1908, he published his book *Energy* (*Die Energie*), which provided the general public a discussion of the significance of the role energy has played from ancient history to the nineteenth century. In his last chapter, he developed his view of 'sociological energetics,' which he developed further in a book he published the next year *The Energetical Foundations of Cultural Studies* (*Energetische Grundlagen der Kulturwissenschaft*) (1909) and other later works including *The Energetical Imperative* (*Der energetische Imperativ*) (1912); *Today's challenges* (*Die Forderung des Tages*) (1910); and *The Philosophy of Values* (*Die Philosophie der Werte*) (1913). He defined cultural progress in terms of a continuing increase in the availability of energy; the efficiency of energy use; and the substitution of human energy by alternative forms. In *The Energetic Imperative*, he developed a scientific system of ethics based on the imperative: "Don't waste energy, make use of it." He did not merely use this imperative as a moral guide; he also used it to understand the course of human history. His work was memorably criticized by Max Weber for applying concepts related to energy to disciplines outside the natural sciences, such as sociology, in ways that were not appropriate (1909), but in recent decades, some scientists have attempted to reclaim Ostwald's significance for contemporary sociology (Martinez-Alier, 1987; Stewart, 2014; Hall & Klitgaard, 2017).

So, by the time Alfred Lotka wrote his paper describing the *Principle of maximum energy flux* (1922), there was already a long tradition of scientists and philosophers that proposed ideas related to this principle, the MPP and the will to power. In the next chapter, we will turn our attention to Lotka's work.

References

Adams, H. (1904a). A law of acceleration. In H. Adams (1919) *The education of Henry Adams*. Houghton Mifflin, ch. 34.

Adams, H. (1904b). A law of acceleration. Edited by Eric Steinhart. From H. Adams (1919) *The education of Henry Adams*. Houghton Mifflin, ch. 34. https://ericsteinhart.com/progress/adams-accelerate.pdf

Adams, H. (1919). *The degradation of the democratic dogma*. The Macmillan Company.

Ansell-Pearson, K. (2009). Free spirits and free thinkers: Nietzsche and Guyau on the future of morality. In J. Metzger (Ed.), *Nietzsche, Nihilism, and the philosophy of the future* (pp. 102–124).

Blumenbach, J. F. (1781). *On the drive for education and the business of procreation* (1st ed.). Johann Christian Dieterich, Göttingen 1781, (online).

Boltzmann, L. (1886). *Der zweite Hauptsatz der mechanischen Wärmetheorie*. Gerold.

Burns, D., & Paton, D. N. (1921). *An introduction to biophysics*. Kessinger.

Drossbach, M. (1884). *Über die scheinbaren und die wirklichen Ursachen des Geschehens in der Welt*. Halle/Saale.

Emden, C. 2014. Nietzsche's naturalism: Philosophy and the life sciences in the nineteenth century, .

Geddes, P. (1881). *The classification of statistics and its results*. A. & C. Black.

Gigantes, D. (2009). *Life: Organic form and romanticism*. Yale University Press.

Guyau, J. 1885 (1921). *Esquisse d'une morale sans obligation ni sanction*. F. Alcan.

Hall, C. A. S., & Klitgaard, K.. (2017). *Energy and the wealth of nations: An introduction to BioPhysical economics* (2nd ed.). Springer.

Heuser-Keßler, M. L. (1986). *Die Produktivität der Natur. Schelling; Naturphilosophie und dis neue Paradigma der Selbstorganisation in den Naturwissenschaften*. Duncker & Humblot.

Hunt, K. (2023). Meet the man who has transformed our understanding of evolution. *CNN*, January 30. https://www.cnn.com/2023/01/30/world/dolph-schluter-profile-crafoord-prize-scn/index.html

Jevons, W. S. (1865). *The coal question: An inquiry concerning the Progress of the nation, and the probable exhaustion of our coal mines*. London: Macmillan and Co.

Kant, I. (1951). *Critique of judgment*. Hafner Press. Translated by Bernard from: Kritik an der Urteilskraft.

Lange, F. (1950). *History of materialism*. 2nd book, trans. E.C. Thomas. Humanities Press. Org. published in German in 1865.

Lotka, A. (1922). Contribution to the energetics of evolution. *Proceedings of the National Academy of Science, 8*(6), 147–151.

Martinez-Alier, J. (1987). *Ecological economics: Energy, environment and society*. Basil Blackwell.

Mayer, J. R. (1845). *Die organische Bewegung in ihrem Zusammenhange mit dem Stoffwechsel: Ein Beitrag zur Naturkunde*. Drechsler.

Mayer, J. R. 1867. Die Mechanik der Wärme in gesammelten Schriften. .

Moore, G. (2002). *Nietzsche, biology and metaphor*. Cambridge. Kindle version.

Nernst, W. (1913). *Theoretische Chemie*. Stuttgart Verlag von Ferdinand Enke.

Nietzsche, F. (1886/1989). *Beyond good and evil* [Jenseits von Gut und Böse. Vorspiel einer Philosophie der Zukunft]. Translated by Walter Kaufmann. Vintage.

Nietzsche, F. (1967). *Werke: Kritische Gesamtausgabe*, founded by Giorgio Colli and Mazzino Montinari, ed. Volker Gerhardt, Norbert Miller, Wolfgang Müller-Lauter, and Karl Pestalozzi (Berlin and New York: Walter de Gruyter, 1967–). The Philologica are quoted according to volume and page number. The Nachlaß is quoted according to volume and fragment number.

Odum, H. T. (1995). Self-organization and maximum power. In C. A. S. Hall (Ed.), *Maximum power: The ideas and applications of H. T. Odum* (pp. 311–330). University of Colorado.

Odum, H. T. (2007). *Environment, power, and society for the 21th century*. Columbia University Press.

Odum, H. T., & Pinkerton, R. (1955). Time's speed regulator: The optimum efficiency for maximum power output in physical and biological systems. *American Scientist., 43*(2), 331–343.

Osborn, H. F. (1918). *2019*. The origin and evolution of life

Ostwald, W. (1908). *Die Energie*. J.A. Barth.

Ostwald, W. (1909). *Energetic foundation of cultural studies [Energetische Grundlagen der Kulturwissenschaft]*. Leipzig, W. Klinkhardt.

Ostwald, W. (1910). *Die Forderung des Tages*. Akademische Verlagsgesellschaft.

Ostwald, W. (1912). *Der energetische Imperativ*. Akademische Verlagsgesellschaft.

Ostwald, W. (1913). *Die Philosophie der Werte*. A. Kröner.

Pfaundler, L. (1902). 'Die Weltwirtschaft im Lichte der Physik'. *Deutsche Revue* (ed. Richard Fleischer)*, 22*(2), April–June, 29–38, 171–82.

Rolph, W. H. (1884). *Biologische Probleme, zugleich als Versuch zur Entwicklung einer rationellen Ethik* (2nd ed.). English.

Sacher, E. (1881). *Grundzuge einer Mechanik der Gesellschaft*. Gustav Fisher.

Schmidt, R. W. (1988). Nietzsches Drossbach-Lektüre: Bemerkungen zum Ursprung des literarischen Projekts "Der Wille zur Macht". *Nietzsche-Studien, 17*.

Schulter, D. (2000). *The ecology of adaptive radiation*. Oxford.

Stack, G. (1983). *Lange and Nietzsche*. de Gruyter.

Steffen, W., Broadgate, W., Deutsch, L., Gaffney, O., & Ludwig, C. (2015). The trajectory of the anthropocene: The great acceleration. *The Anthropocene Review, 2*(1), 81–98.

Stewart, J. (2014). Sociology, culture and energy: The case of Wilhelm Ostwald's 'sociological energetics' – A translation and exposition of a classic text. *Cultural Sociology, 8*(3), 333–350.

Weber, M. (2012 [1909]). 'Energetical' theories of culture. In H. H. Bruun & S. Whimster (Eds.), *Max Weber: Collected methodological essays* (pp. 252–268). Routledge.

Chapter 3
Lotka's Principle of Maximum Energy Flux

Abstract This chapter explores Alfred Lotka's discussion of what he calls the *principle of maximum energy flux*, which H. T. Odum goes on to further develop and refer to as the *maximum power principle*.

3.1 Introduction

When H. T. Odum discusses the maximum power principle (MPP), he usually describes Alfred Lotka as the source of the concept (see, for example, Odum & Pinkerton, 1955, p. 332; Odum, 1971, p. 31, 32; Odum, 2007, p. 37). Lotka (1880–1949) was a physical chemist, biophysicist, mathematician, and statistician who did influential work in demography and energetics. He had a diverse professional career, working as an assistant chemist, a patent examiner, an assistant physicist, an editor for the Scientific American magazine, a staff member at Johns Hopkins University, and a statistician for the Metropolitan Life Insurance Company. He was best known for his work on the Lotka-Volterra predator-prey model he developed independently and at the same time as Vito Volterra; however, he thought his work on energy was more important. In 1922, at the age of 42, Lotka wrote two short papers in which he outlined what he referred to as the *principle of maximum energy flux* (PMEF); in 1925, he briefly described this principle in his book *Elements of Physical Biology*; and, in 1945, he provided a detailed defense of this principle in a paper entitled "The Law of Evolution as a Maximal Principle."

This chapter will provide a description of Lotka's PMEF that follows the trajectory of his own development of the principle: it will discuss the two papers he wrote in 1922 and the paper he wrote in 1945 in the order in which they were published. The focus in Sects. 3.2, 3.3, and 3.4 will be on accurately summarizing Lotka's development of the PMEF. In Sect. 3.5, I will critically analyze a couple of points Lotka makes in these papers that are particularly relevant to our discussion of the MPP.

© The Author(s), under exclusive license to Springer Nature
Switzerland AG 2025
T. McWhirter, *Maximum Power and its Philosophical Roots*, SpringerBriefs in
Energy, https://doi.org/10.1007/978-3-031-80622-3_3

3.2   Contributions to the Energetics of Evolution

Lotka begins his short essay "Contributions to the Energetics of Evolution" (1922a) by noting that Ludwig Boltzmann pointed out that "the fundamental object of contention in the life-struggle, in the evolution of the organic world, is available energy" (Boltzmann, 1886). Lotka notes that Walther Nernst, David Burns, and Henry Fairfield Osborn also viewed evolution from a thermodynamic perspective (Nernst, 1913; Burns & Paton, 1921; Osborn, 1918). All the practices associated with survival and reproduction require the expenditure of available energy, such as finding food, engaging in mate selection and sexual activity, and self-defense. Lotka wrote that "those organisms whose energy-capturing devices are most efficient in directing available energy into channels favorable to the preservation of the species" will have a selective advantage in evolutionary processes (p. 147). Following the lead of Boltzmann and the other scientists he mentions, Lotka reformulates Darwin's theory of natural selection in terms of thermodynamics.

Lotka argues that the first effect of natural selection is that it increases the number or mass of "those organisms whose energy-capturing devices are most efficient in directing available energy into channels favorable to the preservation of the species." He also argues that natural selection has a second effect. If there are sources of available energy in the environment that are in excess of those already being tapped by the entire system of living organisms, those species that have "superior energy-capturing and directing devices" can tap into this "excess" of available energy and "increase the total energy flux through" the system, and "natural selection will operate to preserve and increase" these species whenever they arise. Lotka refers to these two effects of natural selection as the "principle of maximum energy flux": "natural selection tends to make the energy flux through the system a maximum, so far as compatible with the constraints to which the system is subject" (throughout the essay he refers to natural selection operating on both organisms and "natural systems", probably meaning ecosystems in today's language). Lotka explains in a footnote (n. 5) that he uses the term 'energy flux' to "denote the available energy absorbed by and dissipated within the system per unit of time." This demonstrates that his PMEF refers to both the absorption and the dissipation of energy. It also refers to "available energy," which in the contemporary literature is often called *exergy*, and is defined as the amount of work a system can perform when it is brought into equilibrium with its environment.

Lotka explains that the exergy flux through living systems can be increased by either increasing the number of energy transformations or their efficiency: by "enlarging the wheel" or by causing it to "spin faster" (1921). If a human population does not have any more land to use for farming, it can develop an ability to grow crops faster: two crops a year rather than one. This will increase the exergy flux through the system using the same amount of land: it will keep the wheel the same size and just cause it to "spin faster." Another contemporary example using solar energy would involve solar panels. A population that uses solar panels to generate energy can increase the exergy flux through the system by increasing the number of solar panels ("enlarging the wheel"), and/or increasing the efficiency of the solar panels (causing the wheel to "spin faster," i.e., generate more energy from the solar energy captured). Lotka described contemporary human beings as increasing the exergy flux through the living system by *both* enlarging the wheel *and* causing it to spin faster (1922a).[1] He concludes an essay published in 1921 suggesting that "whether, in this [human beings have] been unconsciously fulfilling one of those laws of nature according to which certain quantities tend toward a maximum, is a question well deserving of our attention" (1921). A year later he provides an answer to the question: "this is now made to appear probable; and it is found that the physical quantity in question is of the dimensions of power, or energy per unit of time..." (1922a). In this conclusion, we can begin to see how H. T. Odum might comfortably interpret Lotka's PMEF as a *maximum power principle*.

In an addendum to "Contributions to the Energetics of Evolution," Lotka notes that after he wrote the essay, he received a copy of a book by James Johnstone entitled *The Mechanism of Life*, which draws a different conclusion: "In living processes the increase of entropy is retarded" (Johnstone, 1921, p. 220). This conclusion appears to not be consistent with the PMEF: when the exergy flux through a system is maximized, the dissipation of exergy is maximized, which maximizes the production of entropy.[2] Because of this, Lotka takes some time in the addendum to consider the relation between these two views.

Lotka describes two ways these two views can actually be consistent with each other. First, while some parts of living systems may retard entropy production via storage, the systems, as a whole, may still maximize the exergy flux through them by expending biomass. Plants appear to collect exergy and retard the increase of entropy, as described by Johnstone. Lotka himself suggested that some "living organisms may be capable of retarding the exergy flux through the system of nature" (n. 13). However, Lotka suggests that animals, on the other hand, appear to dissipate exergy and increase the production of entropy.[3] When animals initially emerged in a world filled with plants, it appears that it clearly accelerated the production of entropy of the system as a whole. Therefore, "at certain stages in the evolution of the system, at the least, life must have tended to increase rather than decrease dissipation." Animals remain "essentially a dissipative type, as compared with plants." Lotka contends that in order for Johnstone's argument to apply conclusively to living systems, it would have to show that "the system of coupled transformers, plant and animal, as a whole" evolves in a direction that retards the increase of entropy.

---

[1] Lotka wrote, "The influence of man upon the world's events seems to have been largely to accelerate the circulation of matter and energy through such cycles, either by "enlarging the wheel", i.e., increasing the mass taking part in certain cycles, or else by causing it to "spin faster," i.e., increasing the velocity of the circulation, decreasing the time required for a given mass to complete the cycle" (1921, p. 171–172).

[2] This does not occur at the moment the exergy is stored as biomass.

[3] The biomass of a blue whale would be an exception here.

He does not provide such an argument; therefore, Lotka concludes, it is possible that systems of "coupled transformers" evolve in the direction outlined by the PMEF.[4]

Second, Lotka argues that "where the supply of available energy is limited, the advantage will go to the organism which is most efficient, most economical, in applying to preservative uses such energy as it captures." The value of efficiency will not be as great when the supply of exergy is not limited. In his description of the PMEF, he stipulates that natural selection "tends to make the energy flux through the system a maximum, *so far as compatible with the constraints to which the system is subject*" (my emphasis). This qualification is part of the principle itself. He acknowledges that this qualification "modifies" the principle in a manner that is worthy of "further investigation." As we shall see, Odum and Pinkerton acknowledge the importance of this qualification in their formulation of the maximum power principle. The contemporary scientist Enrico Sciubba suggests that Lotka's acknowledgement of the importance of this qualification is consistent with the "'random emerging, environment-dependent opportunism' that defines our modern vision of evolution" (Sciubba, 2011; Dobzhansky et al., 1977).

Finally, at the end of his essay, Lotka mentions that the PMEF "bears a certain outward resemblance to a principle enunciated by" Wilhelm Ostwald (who we discussed in the previous chapter): "Of all possible energy transformations, that one takes place, which brings about the maximum transformation in a given time" (Ostwald, 1892, p. 37). Lotka suggests that this principle "is based on entirely different grounds from those here brought forward," and "it is not of general applicability, and in particular, its application to systems of the kind here considered does not appear warranted." One can certainly question this claim.

This section provided a picture of Lotka's initial effort to describe his PMEF. His basic theoretical argument for this principle is generally not questioned by later scientists. While some aspects of this principle are surprisingly consistent with contemporary discussions, as Sciubba suggests, other aspects of Lotka's discussion do appear dated and ambiguous, the kind of thing you might expect from a scientist writing over a century ago. However, Lotka's discussion of how the value of efficiency is increased when the available energy is limited goes on to be an important part of the MPP that is often misunderstood by contemporary scientists. The next section takes a look at the second paper he wrote in 1922 and how it relates to his PMEF.

---

[4]A question not addressed by Lotka here is the degree that ecosystems simply change the albedo of the planet. Probably in general ecosystems reduce the reflectance of the rocks, water or glaciers that would otherwise be there alone. The reflected photons might then travel through the universe with high individual exergy and a very low probability of hitting anything that would change their potency, but also to no particular consequence.

3.3 Natural Selection as a Physical Principle

Right after his essay on the energetics of evolution in the *Proceedings of the Nation Academy of Science of the United States*, you find another short article Lotka published that describes natural selection as a physical principle. He begins this essay by stating that the "two fundamental laws of thermodynamics are, of course, insufficient to determine the course of events in a physical system. They tell us that certain things cannot happen, but they do not tell us what does happen" (1922b). According to these laws, the total amount of energy cannot change, although its ability to do work does change, and the entropy of the total amount of energy cannot be reduced. Many scientists in the late 19th and early 20th centuries thought the second law, in particular, was not consistent with the development of organized organic systems described in biological evolution: how do these organized systems develop if entropy cannot be reduced over time? The two fundamental laws of thermodynamics do not provide an answer to this question; Lotka argues that they can if we include among them the principle of natural selection.

Lotka notes that a couple of writers have taken advantage of the explanatory space left by the two fundamental laws of thermodynamics and have described the effect that life can have on physical systems. He mentions how Wilhelm Ostwald observes that "the organism utilizes, in manyfold ways, the freedom of choice among reaction velocities, through the influence of catalytic substances, to satisfy advantageously its energy requirements" (1902, p. 328). Sir Oliver Lodge "has drawn attention to the guidance exercised by life and mind upon physical events, within the limits imposed by the requirements of available energy" (1906, p. 144). Hyacinthe Guilleminot "sees the influence of life upon physical systems in the substitution of guidance by choice in place of fortuitous happenings, where Carnot's principle leaves the course of events indeterminate" (Lotka, 1922b, p. 151; Guilleminot, 1919, p. 121). Lotka takes issue with Guilleminot, arguing that fortuitousness is "not an objective quality of a given event. It is the expression of our subjective ignorance, our lack of complete information, or else our deliberate ignoring of some of the factors that actually do determine the course of events."

Lotka argues that just because the fundamental laws of thermodynamics do not tell how physical systems develop over time does not necessarily mean that this development is "actually indeterminate, ... It merely means that the laws, as formulated, take account of certain factors only, leaving others out of consideration; and that the data thus furnished are insufficient to yield an unambiguous answer to our enquiry regarding the course of events in a physical system" (1922b, p. 152). He argues that whether we are dealing with biotic or abiotic systems, the development of these systems is "determinate, though not in terms of the first and second laws alone." The "freedom" of which systems "avail themselves under the laws of thermodynamics" is a "spurious" one "arising out of an incomplete statement of the physical laws applicable to the case."

Lotka argues that this "spurious freedom" emerges from all systems that receive exergy from their environment and use it to develop. These systems are now referred to as developing "out of equilibrium." When systems are moving toward an equilibrium state—"such as thermally and mechanically isolated systems"—"the first and second laws of thermodynamics suffice" to determine the end state. But when systems receive a steady supply of exergy—"such as the earth illuminated by the sun"—and are moving toward a "stationary state" rather than a "true equilibrium," the first and second laws are not sufficient to determine the end state. These systems that develop out of equilibrium can include constituents that are "auto-catalytic or auto-catakinetic" and natural selection can operate upon them.[5] He writes that "such auto-catakinetic constituents are the living organisms, and to them, therefore the principles here discussed, apply" (*i.e.*, the principles of natural selection and maximum exergy flux) (1922b, p. 153).[6]

Lotka provides a view of organic evolution from the perspective of the fundamental thermodynamic laws, with the addition of the principle of natural selection, which he describes as a "physical principle." He writes that the "battle array" of evolving systems is "presented to our view as an assembly of armies of energy transformers," "armies composed of multitudes of similar units," which "range themselves according to law and order, for those species of units, those types of transformers, are picked out for survival, whose mechanism possesses certain definite properties. Thus the principle of natural selection makes its entry into dynamics." Lotka acknowledges that Guilleminot had already argued that the "principle of natural selection is competent to yield information beyond the scope of the laws of thermodynamics." Lotka concludes that it functions "as a third law of thermodynamics (or a fourth, if the third place be given to the Nernst principle)."

This second paper from 1922 adds to our understanding of Lotka's PMEF. He believed this principle guides natural selection and, in this paper, he argues that natural selection applies to all thermodynamic systems developing out of equilibrium and should be considered a fundamental law of thermodynamics. One might think that this would lead Lotka to describe his PMEF as applying to all natural systems, biotic and abiotic. However, as we will see in the next section, this is not the path he decides to take 23 years later.

3.4 The Law of Evolution as a Maximal Principle

In 1945, four years before his death, Lotka published a long essay entitled "The Law of Evolution as a Maximal Principle," which provides the most comprehensive argument that the PMEF guides evolution. He begins this essay acknowledging that fundamental to the concept of evolution is the idea that it is *directed*. The objective of the essay is to define as clearly as possible how evolution is directed. He

---

[5] Auto-catalytic or auto-catakinetic constituents are those that cause their own chemical reactions.

[6] Lotka notes that D'Arcy Thompson attributes the origination of this idea to Chodat, quoted by Monnier, A. (Thompson, 1917, p. 132).

considers a number of possible ways evolution has been described as being directed and discusses the problems with them; then, he describes his PMEF and provides arguments that support it.

Lotka notes that some have "said or implied, that the direction of evolution is the direction of progress…;" however, he argues that this is "merely to substitute for an undefined term another at best ill defined, and contaminated with anthropomorphic flavor." The same objection can be made to attempts to describe evolution as a "passage from lower to higher forms." If we describe evolution as leading in the direction of greater complexity, then, Lotka suggests, this direction is "poorly defined" because there are exceptions to this rule and we are looking for a "law of nature that brooks no exceptions."

One description of the direction of evolution that has come to be a cornerstone of modern statistical thermodynamics is that evolution leads in the direction that is most probable. If, for example, the observable universe starts from a state of low entropy, it will evolve in a direction that increases overall entropy because that is more probable. However, Lotka argues that the idea that evolution leads in the direction that is most probable is not precise. He illustrates this problem using an example that involves shuffling a new deck of cards. Let's say we have a new deck of cards where the first 13 cards "are the Ace to Ten, Knave, Queen, King, of Spades, arranged in that order." If we shuffle the cards thoroughly, the spades will be spread "approximately evenly throughout" the deck. One could say that shuffling the cards moved the deck "from a less probable to a more probable state." Lotka argues though that this idea of a "more probable state" is not "an objective property of either the pack or the process of shuffling. It is a subjective property of the shuffler, or the observer, who takes a special interest in the suit of spades picked out for observation…" The process of shuffling the deck of cards did not take it "from a less probable to a more probable state, but from a precisely known to a less well-known state." Viewed objectively, the "final state is just as probable or improbable as the initial state. The distinction is not objective but psychological." The statement that evolution leads in the direction of what is most probable is, "not so much a false statement, as it is a statement devoid of meaning, unless it is further supplemented by a specification as to how and in what respects the probabilities of" the more or less probable states are "measured."

Lotka next considers the idea that evolution is an irreversible process. He explains that "irreversible" processes have a specific meaning for physicists. Once the new deck of cards has been shuffled, it will take forever to return the deck to its original order by merely re-shuffling it. In that sense, shuffling the cards is an irreversible process, but this is not the kind of "irreversibility" intended by physicists. They use the term to refer to processes "associated with [a] decrease in thermodynamic potential, with [a] capacity for yielding a balance of work." For example, when a tire runs over a nail that punctures it, the air will begin to leak out of the tire; the air will not go into the tire, filling it up. When the air leaks out of the tire, it will no longer be able to do the work of inflating the tire and enabling the car to go down the road. When a glass is dropped on the kitchen floor, it will break into a bunch of different pieces; if these pieces are dropped on the kitchen floor, they will not magically form a new glass. When the glass is broken, it will no longer be able to do the work of holding water. These are the kind of processes physicists are referring to when they discuss "irreversible" processes.

Lotka explains that the irreversible processes to which physicists refer play a role in biological evolution at different scales; however, to state that "a biological species never retraces its steps," or that "when a race has lived its term it comes no more again," does not "characterize an 'irreversible' process in a useful sense, for this does not distinguish the chain of events from a mere 'changeful sequence.'" If the course of events that occur through an evolutionary process are a "changeful sequence," then they could occur in reverse order. Lotka concludes that the kind of "irreversibility" involved in evolutionary processes "has no prognostic value:" "It tells us at best that if sequence A, B, C happens, then the sequence C, B, A will not happen and vice versa; but it does not tell us which of the two actually will happen." Here, I am merely describing Lotka's argument; I am not taking a position on it, one way or another.

One last proposed direction for evolutionary processes is considered by Lotka: things evolve in a direction that increases their organization. His first concern with this proposal is that no one has developed a way to measure organization so we do not have a way to empirically test the proposal. His second concern is that the proposal belongs in the "domain of descriptive science." He maintains that we are seeking to go beyond this domain and develop "a deductive scheme in which the law of evolution is seen to flow as a necessary consequence from fundamental laws, as a result of the physical properties of the organisms and the system of which they form a part."[7] He argues that the idea that increasing organization is a "common characteristic of organic evolution" is "far too complicated and too inexactly defined a conception to be classed with such basic principles as the two fundamental laws, for example." If we were to accept this idea as a basic principle, we may discourage further investigation that could uncover a proposal that outlines a direction for evolution that is based on "*necessary* relations deduced from known universal principles."

So, Lotka does not accept these proposals for what directs evolutionary processes: progress, greater complexity, greater probability, irreversibility, or increasing organization. His first suggestion of a "signpost" that leads us in a more helpful direction is surprising. He notes that the philosopher Bertrand Russell described living things as "imperialists" that seek to transform into themselves as much of their environment as possible. Lotka thought all of evolution flowed from this "'chemical imperialism' of living matter" (1927, p. 27). We can already see how this kind of 'chemical imperialism' can lead us to the PMEF. It also leads us to recognize that, as Lotka writes, the "the problem of organic evolution, as a problem in the distribution of matter among the components of a material system, is formally in the same general category as these problems of distribution of physico-chemical

---

[7]Notice how Lotka's view here stands in contrast with the view of Mayer, who believed all he could do was develop equations that accurately described different energy transformations; he could not explain the physical processes that caused them (Mayer, 1978, p. 10).

systems..." (1945, p. 176). Lotka acknowledges that there are, of course, differences between organic and physico-chemical systems. He outlines four of them.

First, the "individual components recognized in physico-chemical transformation–molecules and atoms–escape direct observation by any of the ordinary methods." We have the ability to observe the "bulk properties and effects" of molecules and atoms, "such as volume, pressure, temperature, etc.," but the investigation of the behavior of individual molecules and atoms "requires altogether different, highly refined techniques." When we investigate organic systems, the situation is "diametrically opposite." In this case, the individual components are the organism, and they can be observed directly; the collective behavior of these organisms "*as a whole*," on the other hand, require the "development of a special branch of statistical dynamics."

Second, "physico-chemical transformations (exclusive of transmutations of elements) are bound by certain equations of constraint (the "reaction equations") which are fixed for all time. A system composed of hydrogen and oxygen, for example, may give rise to different quantities of H_2, O_2, and H_2O, but it will never create hydrochloric acid. Lotka explains that the transformations that involve organic species are "of a different order." They are transformations of "*structure*" that are, in many ways, "capable of infinite variation." Over long periods of time, the evolutionary competition between species gives rise to new species, which, in many respects, are very different. Lotka describes the "equations of constraint" for organic transformations as not only being more liberal than those of physico-chemical transformations, but they are "functions of time, whereas the physico-chemical equations of constraint (*e.g.*, the "reaction equations" of chemistry) are rigidly fixed for all time."

This second difference is closely related to the third: the components of physico-chemical systems are homogeneous; the components of organic systems, on the other hand, are "only relatively homogeneous." The evolution of organic systems occurs among different species—*inter-species evolution*—and within each species—*intra-species evolution*. The variations that emerge through inter-species evolution are far more extreme than those we find in intra-species evolution; however, both forms of evolution illustrate how organic systems are only "relatively homogeneous." Because the components of physico-chemical systems are homogeneous, there is no intra-species evolution.

The fourth difference between organic and physico-chemical systems is that the transformations of physico-chemical systems are "usually considered" as "taking place under conditions leading to an *equilibrium*." The evolution of organic systems, on the other hand, is understood as taking place while these systems are developing out of equilibrium, *i.e.*, while they are receiving a flow of exergy from their environment. They are not viewed as headed toward an equilibrium; they are viewed as headed toward a "stationary state" that is sustained by the flow of exergy from their environment.

Lotka suggests that the evolution of organic nature is best understood as involving "aggregates of energy transformers adapted by their composition and structure to guide available energy" into those channels that enhance their "maintenance and growth." In order to better understand this evolution, a "special branch of physics needs to be developed, the *statistical dynamics of systems of energy transformations*." He notes that organic systems, like inorganic systems, are subject to physical laws, like the first law of thermodynamics; so, this special branch of physics will include physical laws. The transformers (e.g., organisms) considered in this branch of physics will include an important class that acquires its exergy from "sources distributed discontinuously" throughout the environment. This requires these transformers to have structures that enable them to effectively locate their sources of exergy. The more effective these structures are, the more efficient these transformers will be, *i.e.*, they will be able to spend less exergy acquiring exergy.

Lotka discusses four typical types of structures. First, there are *"Receptors."* These structures represent or depict the environment to the transformer. Eyes are an example of receptors. The second is *"Free Energy."* The transformer must have free energy stored that can be directed to channels that will enable it to acquire exergy from the environment when it is located. The third type of structure is *"Effectors."* These structures enable a transformer to "bring about an encounter with an energy source" once one is detected by a receptor and to avoid being captured, injured, or killed. Finally, the fourth type of structure is *"Adjusters."* These structures enable the effectors to be adjusted or adapted in accordance with the information provided by the receptors.

This discussion of these four types of structures has presupposed that there are exergy sources located in different areas of the environment. In many cases, these exergy sources are another type of energy transformer, one that is sustained by an exergy source that is "continuously distributed in space," like sunlight or "dissolved substances" that drift in the water. Plants would be an example. These transformers can remain stationary and passive.

The special branch of physics that Lotka is proposing—the *statistical dynamics of systems of energy transformation*—would focus on the data outlined above. While he describes the "energy transformers" as "'living' animals and plants," he argues that it "will be desirable to develop the discipline of statistical dynamics of aggregates of transformers *irrespective of this fact*." (p. 182). If we do so, this discipline will apply to "the world of organisms in so far as they are energy transformers having the physical properties taken in view in that discipline, just as the law of conservation of energy, is applicable to them:" it will provide a view of life from the perspective of thermodynamics. Lotka argues that this approach provides some important advantages. First, we "have complete control of the particular properties which we may assign to the transformers under consideration, without raising the question whether such a transformer actually exists or ever has existed." We have the ability to "isolate the factors essential to our enquiry." This is particularly important for a new scientific discipline, like the one he envisions: he argues that the history of science makes this clear (n. 22). Here, Lotka is suggesting that the factors essential to our inquiry into evolution are thermodynamic.

One issue that is central to Lotka's inquiry into evolution is the relation between his PMEF and Johnstone's principle that "in living processes the increase of entropy is retarded" (1921). He had briefly addressed this issue in the addendum to his essay "Contribution to the Energetics of Evolution," written 23 years earlier, suggesting that there were two ways the principles could work together. The first way outlined there is described in more detail in "The Law of Evolution as a Maximal Principle." He suggests that if we assume that the evolution of life is based solely on his principle or Johnstone's, a prima facie review of these two opposing principles can be perplexing; if, on the other hand, we acknowledge that they both work together, "going on side by side," the "perplexity vanishes."

He provides a simile to illustrate how the two principles work "side by side." Imagine a reservoir that is used to gather rainwater, like what are now referred to as rain gardens. One part of the rain garden guides water into the garden. It can be enlarged or reduced to increase or decrease that amount of water that flows into the garden. A second part of the garden allows water to drain from the garden. It can be enlarged or reduced to increase or decrease the amount of water that drains from the garden. In one sense these two parts of the garden appear to be opposed to each other, or working against each other; however, in another sense they work together: *"together they increase the total flow through the reservoir."* Generally, plants gather the exergy and animals dissipate it: together, they increase the total flow of exergy through living systems. This passage demonstrates more clearly that Lotka believed that his principle and Johnstone's worked together in a way that is consistent with his principle. Lotka uses this example to set the stage for the argument he provides for his PMEF in his 1922 essay (1922a). He provides a two-page quote from that essay. At the age of 65, he still stood by the argument he made when he was 42.

In Sect. 3.5, I critically analyze a couple aspects of Lotka's work from the perspective of contemporary science. Understandably, some of his views do not hold up very well, but in many cases his ideas are surprisingly prescient, such as his view, mentioned earlier, that the meaning of the PMEF is modified by the fact that it applies to natural systems in a manner that is "compatible with the constraints to which the system is subject" (1922a).

3.4.1 Human Behavior and the Principle of Maximum Energy Flux

At the end of "The Law of Evolution as a Maximal Principle," Lotka provides a brief analysis of human behavior from the perspective of his PMEF. He begins this section noting that human evolution "in more recent times" has followed an "entirely new path" that transcended the limitation of genetics: "In place of slow adaptation of anatomical structure and physiological function in successive generations by selection survival, increased adaptation has been achieved by the incomparably more rapid development of "artificial" aids to our native receptor-effector apparatus, in a process that might be termed *exosomatic* evolution." In another essay, he provides a chart of the "rocket-like ascent, in modern times, of human knowledge and technical skill", which is based on the number of pages in *Darmstaedter*—the Handbook of the History of Natural Science and Technology—devoted to scientific discoveries and inventions over the centuries, from 1001 to 1900 (Lotka, 1939, p. 625, 1945, n. 32). The graph shows a nonlinear increase in discoveries and inventions that continues to accelerate (1939, p. 625) (Fig. 3.1).

Exosomatic evolution is the term he uses to refer to the process through which human knowledge and technical skill is developed and passed onto future generations. In contemporary science, this process is referred to as *cultural evolution*, the study of which was initiated in the 1970s and early 1980s by Cavalli-Sforza and Feldman (1981) and Boyd and Richerson (1985, 2005). Lotka's discussion of this process in 1945 is particularly prescient.

Lotka suggests that "until recently" the net effect of exosomatic evolution has been consistent with the PMEF. "By ingenious contrivances [human beings have] immensely refined and multiplied the operation of [their] receptor-effector apparatus." As a result, humans have completely transformed their "methods of production" in what is now referred to as the agricultural and industrial revolutions.

Lotka argues, however, that this is only "one side of the picture." He wrote this essay during World War II, while the "human knowledge and technical skill" passed down through the generations was being used to develop unique weapons of war and kill an unprecedented number of people. Exosomatic evolution led to the decline of the death rate, but it also led to a world at war where, he writes, "contending armies exchange death against death." Massive investments in military technology and resources were made while occupants of occupied countries were forced to use the majority of their remaining resources to satisfy, as best they could, their most basic needs.

Exosomatic evolution also led to other problems, Lotka explains. The dramatic increase in the efficiency of production made it possible to provide the basic resources needed for survival using a fraction of the workers that used to be required. To secure full employment, a larger portion of industrial production had to be shifted to manufacturing luxury items. While the demand for the necessities of life generally remains relatively stable, this is not always the case for luxuries: the demand for them can wax and wane. This can cause economic instability, such as that witnessed during the great depression, where Lotka writes there was "the incongruous spectacle that in the midst of plenty there·is widespread want."

Lotka suggests that it is quite possible that over time adjustments can be made that reduce or eliminate "at least the more extreme fluctuations" in economic inequality; however, the shift to producing more and more luxury items leads to another more important concern for Lotka. He writes that "few persons may be tempted to go hungry in order to buy jewels. But many actually do go without children, or without an adequate number of them, in order to maintain certain standards of living." He provides a table of the net reproduction rate of different industrial societies between 1931 and 1941 (1945, p. 191). This rate is "the ratio of total births in successive generations." For many countries during this time, this rate was "well below unity, the required minimum" for a society to maintain its size (Fig. 3.2). Lotka warns that "the meaning of these figures is that certain contingents of mankind are headed for extinction, if present attitudes continue."

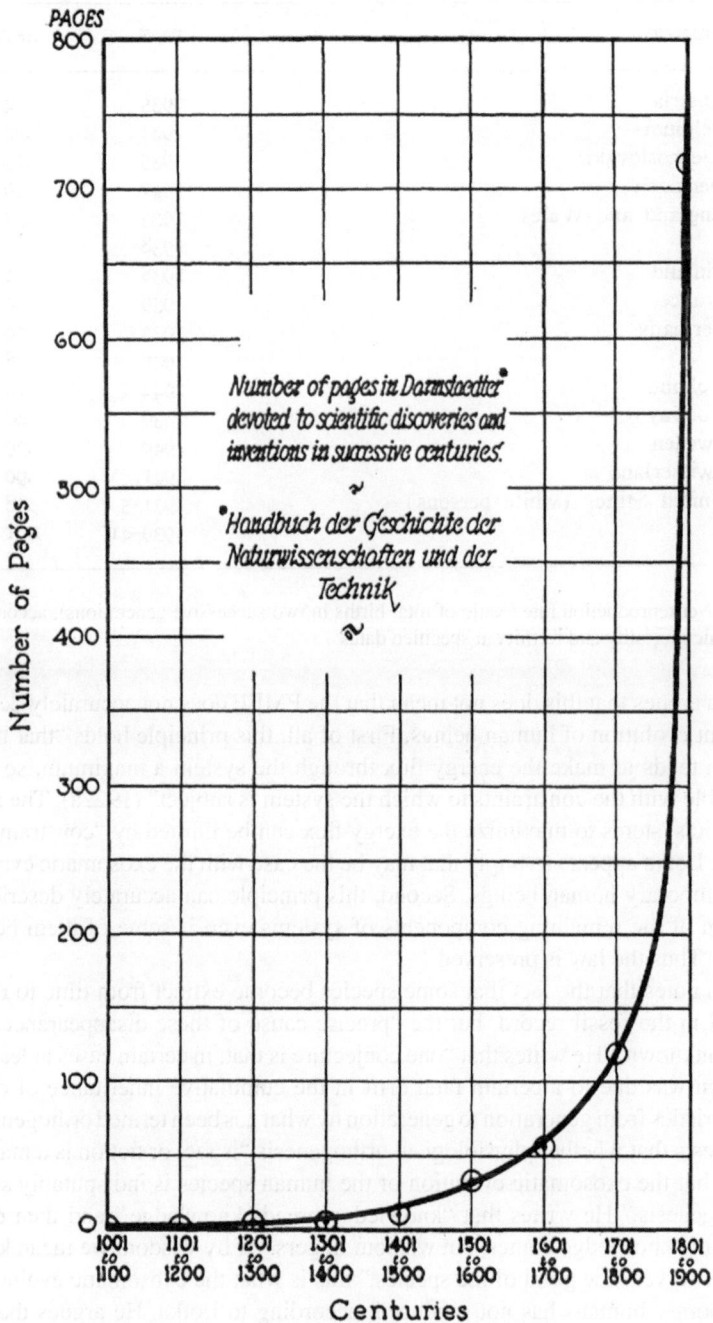

Fig. 3.1 Lotka's growth curve of human knowledge

COUNTRY	DATE	RATE
Austria	1935	.64
Belgium	1941	.67
Czechoslovakia	1935	.79
Denmark	1941	.96
England and Wales	1933	.73
	1938	.81
Finland	1938	.96
France	1939	.90
Germany	1933	.70
	1940	.98
Holland	1941	1.16
Norway	1939	.86
Sweden	1940	.79
Switzerland	1941	.90
United States (white persons)	1931–5	.98
	1939–41	1.01

Fig. 3.2 Net reproduction rate (Ratio of total births in two successive generations), according to age-specific mortality and fertility at specified dates

Lotka argues that this does not mean that the PMEF does not accurately describe the recent evolution of human beings. First of all, this principle holds "that natural selection tends to make the energy flux through the system a maximum, so far as compatible with the constraints to which the system is subject" (1922a). The ability of organic systems to maximize the exergy flux can be limited by "constraints" on systems. Lotka appears to imply that may be the case with the exosomatic evolution of contemporary human beings. Second, this principle can accurately describe the evolution of the remaining components of systems even if some of them become extinct: "Thus the law is preserved."

Lotka notes that the fact that some species become extinct from time to time is reflected in the fossil record, but the "precise cause of these disappearances may remain unknown." He writes that "one conjecture is that, in certain cases at least, the extinction was due to a certain fatal drift in the cumulative inheritance of certain characteristics from generation to generation by what has been termed orthogenesis."[8] He suggests that whether physiological orthogenesis "is fact or fiction is a matter of dispute, but the exosomatic evolution of the human species is indisputably subject to orthogenesis." He writes that "knowledge breeds knowledge," and then quotes Tennyson: "knowledge comes, but wisdom lingers." If by wisdom we mean knowledge that serves "the good of the species," this is what the exosomatic evolution of contemporary humans has not produced, according to Lotka. He argues that "the

---

[8] Orthogenesis is a now obsolete hypothesis that organisms have an innate tendency to evolve in a teleological manner, towards some goal.

receptors and effectors have been perfected to a nicety; but the development of the adjusters has lagged so far behind, that the resultant of our efforts has actually been reversed. From the preservation of life we have turned to the destruction of life.... If certain contingents of the human species are to be preserved and perpetuated, a revision of prevailing valuations will be necessary." As Lotka approaches his conclusion, he provides a quote from the biologist Ralph Lillie: "the avoidance of useless conflict, and the subordination of individual interests to the integrated whole, which includes the individual, would seem to be a rational aim for conscious beings" (1943).

This section has described the argument Lotka provides in "The Law of Evolution as a Maximal Principle" that evolution is directed by the PMEF. Lotka spent a considerable amount of time arguing that other principles do not guide evolution. His argument that the PMEF guides evolution is based primarily on the theoretical argument he makes in his 1922 paper (1922a). In this 1945 essay, he also critically analyzes human behavior from the perspective of the PMEF, which is something we see further developed by Odum and Nietzsche. The next section critically analyzes two aspects of Lotka's description of the PMEF: its relation to exosomatic evolution and abiotic systems.

3.5 Critical Analysis

3.5.1 Exosomatic Evolution and the PMEF

In many respects, Lotka's work is extraordinarily prescient, and, as we will see, it has had a considerable impact. However, like any other human being, including those that point toward new ways of understanding things, his work can benefit from a critical analysis that seeks to better understand what he has written within the context of his other texts and subsequent developments in the sciences. That is the aim of this section.

One of the first things that jumped out at me when I read Lotka's essay "The Law of Evolution as a Maximal Principle" was his suggestion that the exosomatic evolution of contemporary human beings was, in some cases, not consistent with his PMEF. Then, on top of this, he suspects this inconsistency may be the result of the "fact" that this exosomatic evolution is teleological and "indisputably subject to orthogenesis." Orthogenesis is a now obsolete biological hypothesis that living organisms have an innate tendency to evolve in a definite direction. This view was advocated by Jean-Baptiste Lamarck and Henri Bergson, among others. I suspect that Lotka's reference to orthogenesis will get the attention of many knowledgeable readers. This critical analysis will therefore begin with this aspect of Lotka's position.

Lotka's description of human exosomatic evolution was published thirty years before scientists began to systematically study human cultural evolution. It is understandable that he did not fully consider the significance of exosomatic exergy dissipation, *i.e.*, when humans started expending exergy outside their bodies through the use of fire and machines. The Russian biophysicist Aleksandr Il'ich Zotin has been described as producing studies that provide a "good illustration and extension of Lotka's principle" of maximum exergy flux (Martyushev & Seleznev, 2006; Zotin & Zotina, 1993; Zotin & Zotin, 1996, 1999; Zotin et al., 2001). He argues that as bodies expend more exergy, they produce more heat. If this heat is not controlled, it can cause the body to break down. Nature handled this initially through heat regulation mechanisms, like sweating, but eventually humans began to expend exergy outside their bodies, exosomatically, and this enabled them to overcome the thermodynamic limitations of their bodies and dramatically increase the exergy flux of human social systems through the use of fire, machines, engines, trains, cars, refrigerators etc. The development of exosomatic dissipation is, therefore, viewed by Zotin as empirical evidence that is consistent with the PMEF.

The industrial revolutions dramatically enhanced the exosomatic dissipation of exergy, and, as a result, the PMEF came to be less dependent on population growth in human societies. As described in demographic transition theory, when societies become industrialized, their population growth begins to decrease; however, the people in these societies begin to expend far more exergy outside their bodies, so, the use of exergy in these societies continues to go up in a non-linear fashion (Demeny et al., 2003). Lotka notes that in industrial societies that have shifted to the production of more luxuries, the appetite people have "for automobiles, radios, fur coats, jewelry, actually seems to follow the rule of the French proverb *l'appetit vient en mangeant* (appetite comes with eating)" (1945). Consequently, the data Lotka provides on the decrease of the net reproduction rate in the industrial societies around the world between 1931 and 1941 is actually consistent with the PMEF: it is a sign of the extent to which the flux of exergy in industrial societies has shifted from a dependence on population growth to a dependence on the exosomatic dissipation of exergy.

Lotka wrote his 1945 essay during World War II. This would have to have an effect on anyone evaluating the contemporary evolution of human beings. The carnage, brutality, the waste of resources, the inability to work together would understandably be foremost in one's mind during this time. This could clearly make it more difficult to see a couple of hard truths. First, in many countries, World War II brought the Great Depression to an end, dramatically increasing employment and the use of exergy worldwide. It takes a lot of exergy to build the planes, tanks, ships, and weapons and use them in battle. On top of all this, the war led to the creation of atomic energy. Lotka writes in a footnote to his discussion of WWII that "the latest development in this field is the unlocking of atomic energy for purposes of war. It remains to be seen whether this new channel for the flux of energy through the evolving system of which human society form a part, will in the course of time be put to constructive use. In that case this will be the superlative example of the principle of maximum energy flux as characterizing the direction of evolution" (1945). Atomic energy has indeed been put to constructive use in countries around the world.

Lotka's suggestion that WWII and the decrease in the net reproduction rate in industrial societies in the twentieth century demonstrate that the exosomatic evolution of human beings during this period was inconsistent with the PMEF principle reflects a failure to appreciate the extent to which the rapidly increasing exosomatic expenditure of exergy separated the flux of exergy through social systems from a dependence on population growth. When we acknowledge this fundamental shift in the flux of exergy, we can see that the exosomatic evolution of human beings during this period is, in fact, consistent with the PMEF. We, therefore, have no need to appeal to the now obsolete biological hypothesis of orthogenesis to explain the exosomatic evolution of human beings during the twentieth century.

3.5.2 Abiotic Systems and the PMEF

Lotka writes that "the transformers in nature" that are the subjects of his "statistical dynamics" of evolving systems "are 'living' animals and plants" (1945). This may have caught some knowledgeable readers by surprise. In his earlier essay "Natural Selection as a Physical Principle," Lotka goes out of his way to describe the systems that evolve in thermodynamic terms, as "*energy transformers.*" He argues that "in systems evolving toward a true equilibrium," the fundamental laws of thermodynamics can be used to determine their end state; however, when systems are developing out of equilibrium, whether they are biotic or abiotic, the "laws of thermodynamics are no longer sufficient to determine the end state:" the principles of natural selection and maximum energy flux are required. He, therefore, describes the PMEF applying to the evolution of systems out of equilibrium and he suggests "the auto-catalytic and the auto-catakinetic constituents of these systems are the living organisms" (1922b). When he writes that "the transformers in nature are 'living' animals and plants," he seems to be suggesting the PMEF applies *only* to the evolution of organic systems.

In his 1945 essay, Lotka describes four differences between organic and "physico-chemical systems:" organic systems can be observed directly, physico-chemical systems cannot; there are constraints on the transformations of physico-chemical systems that do not apply to organic systems; physico-chemical systems are more homogenous than organic systems; physico-chemical systems are "usually considered" as "taking place under conditions leading to an *equilibrium,*" while organic systems are understood as developing out of equilibrium. Lotka appears to use these four differences to explain why the PMEF does not apply to physico-chemical systems.

Here, Lotka appears to overlook the fact that there are physico-chemical systems that develop out of equilibrium, such as hurricanes, ocean currents (Kleidon, 2009), and streams (Langbein & Leopold, 1962). Lotka cites a passage from the work of D'Arcy Thompson where he writes: "the phenomenon of autocatalysis is by no means confined to living or protoplasmic chemistry" although it is constantly associated with organic systems (1922b; Thompson, 1917). H. T. Odum saw the MPP applying to physical systems: hurricanes, streams, and stars (1994, 1995). Contemporary scientists, such as Enrico Sciubba, suggest that living organisms should be considered as examples of "very complex, auto-catalytic systems" (2011). If we accept Lotka's description in "Natural Selection as a Physical Principle" (1922a) of the significance of the difference between systems developing toward equilibrium and those developing out of equilibrium, we must acknowledge that the PMEF applies to systems that develop out of equilibrium, including those that are abiotic.

3.6   Conclusion

In this chapter, we reviewed the three papers in which Lotka describes his PMEF. Sciubba notes that at the end of "Natural Selection as a Physical Principle" Lotka writes that he has additional material on the "dynamics of evolution" that will be brought forward at another time, but Sciubba claims that he never published this material (2011). Sciubba overlooked Lotka's 27 page paper that he wrote near the end of his life, "The Law of Evolution as a Maximal Principle": his most detailed argument for the PMEF. Instead, he considered only "Contributions to the Energetics of Evolution" and "Natural Selection as a Physical Principle," which are both three and a half pages long. Thus, Sciubba's critical review of Lotka's PMEF—entitled "What did Lotka really say? A critical reassessment of the 'maximum power principle' –considered only approximately ¼ of what Lotka had to say about this principle. We have been able to review a far more comprehensive presentation of the PMEF.

Later in his life, Lotka was able to clarify his position on certain issues, including his view of the relation between his PMEF and Johnstone's principle. We also found more evidence of his prescience, as, for example, illustrated in his discussion of exosomatic evolution. Some problems did emerge in Lotka's position regarding the nature of the systems to which his PMEF applies and its relation to contemporary exosomatic evolution. Thus, we were able to see a more complete picture of Lotka's PMEF, warts and all.

Lotka formulates his PMEF based on theoretical considerations related to his reading of Boltzmann and other scientists.[9] He does not provide a great deal of empirical evidence to support this principle. There was not a lot of evidence available at the time. But things have changed: as we shall see in the next chapter, we now have a wealth of evidence that is consistent with his principle.

---

[9] Sciubba suggests that Lotka's formulation of the PMEF was presented as an "empirical observation" because Lotka's writes that "in every instance considered, natural selection will … increase the total energy flux through the system" (Sciubba, 2011, p. 1348; Lotka, 1922a, p. 148). But Lotka does not say that he considered these instances empirically, and he provides no evidence that he did so; thus we are left with the conclusion that he considered these instances theoretically.

In the next chapter, we consider the work of the prolific systems ecologists who further developed Lotka's PMEF: H. T. Odum.

References

Boltzmann, L. (1886). The second law of thermodynamics. In B. McGinness (Ed.), *Ludwig Boltzmann: Theoretical physics and philosophical problems: Selected writings* (pp. 14–32). D. Reidel, 1974.

Boyd, R., & Richerson, P. J. (1985). *Culture and the evolutionary process*. University of Chicago Press.

Burns, D., & Paton, D. N. (1921). *An introduction to biophysics*. Kessinger.

Cavalli-Sforza, L. L., & Feldman, M. W. (1981). *Cultural transmission and evolution: A quantitative approach (Monographs in population biology 16)*. Princeton University Press.

Demeny, P., McNicoll, G., & Hodgson, D. (2003). "Warren Thompson". *Encyclopedia of population*, Vol. 2. Macmillan reference (pp. 939–940). ISBN 978-0-02-865677-9.

Dobzhansky, T., Ayala, F. J., Stebbins, G. L., & Valentine, J. W. (1977). *Evolution*. W.H. Freeman.

Guilleminot, H. (1919). *La matiere et la vie*. Flammarion.

Johnstone, J. (1921). *The mechanism of life*. Edward Arnold.

Kleidon, A. (2009). Nonequilibrium thermodynamics and maximum entropy production in the earth system: Applications and implications. *Naturwissenschaften, 96*, 653–677.

Langbein, W. A., & Leopold, B. (1962). *The concept of entropy in landscape evolution*. U. S. Geological Survey Prof. Paper 550A.

Lillie, R. S. (1943). The psychic factors in living organisms. *Philosophy of Science, 10*, 262–270.

Lodge, S. O. (1906). *Life and matter*. Williams & Norgate.

Lotka, A. (1921). Note on moving equilibra. *Proceedings of the National Academy of Sciences of the United States of America, 7*(6), 168–172.

Lotka, A. (1922a). Contribution to the energetics of evolution. *Proceedings of the National Academy of Science, 8*(6), 147–151.

Lotka, A. (1922b). Natural selection as a physical principle. *Proceedings of the National Academy of Science, 8*(6), 151–154.

Lotka, A. (1939). Contact points of population study with related branches of science. *Proceedings of the American Philosophical Society, 80*(4), 601–626.

Lotka, A. (1945). The law of evolution as a maximal principle. *Human Biology, 17*(3), 167–194.

Martyushev, L. M., & Seleznev, V. D. (2006). Maximum entropy production principle in physics, chemistry and biology. *Physics Reports, 406*(1), 1–45.

Mayer, J. R. (1978). *Die Mechanik der Wärme: Sämtliche Schriften*. H. P. Münzenmayer e Stadtarchiv Heilbronn Eds. Stadtarchiv Heilbronn.

Nernst, W. (1913). *Theoretische Chemie*. Stuttgart Verlag Von Ferdinand Enke.

Odum, H. T., & Pinkerton, R. (1955). Time's speed regulator: The optimum efficiency for maximum power output in physical and biological systems. *American Scientist, 43*(2), 331–343.

Odum, H. T. (1971). *Environment, power, and society*. Wiley.

Odum, H. T. (1994). *Ecological and general systems: An introduction to systems ecology*. University Press of Colorado.

Odum, H. T. (1995). Self-organization and maximum empower. In C. A. S. Hall (Ed.), *Maximum power: The ideas and applications of H. T. Odum*. University Press of Colorado.

Odum, H. T. (2007). *Environment, power, and society for the 21th century*. Columbia University Press.

Osborn, H. F. (1918/2019). *The origin and evolution of life*. Wentworth.

Ostwald, W. (1892). *Lehrbuch der allgemeinen Chemie* (Vol. 2). Engelmann.

Ostwald, W. (1902). *Vorlesungen ilber Naturphilosophie. Gehalten im Sommer 1901 and der Universitat Leipzig. 2. Auflage.* Leipzig, Veit.

Russell, B. (1927). *An outline of philosophy.* London: George Allen and Unwin.

Sciubba, E. (2011). What did Lotka really say? A critical reassessment of the "maximum power principle". *Ecological Modeling, 222,* 1348–1353.

Thompson, D'. A. (1917). *On growth and form.* Cambridge University Press.

Zotin, A. A., & Zotin, A. I. (1996). Thermodynamic bases for developmental processes. *Journal of Non-Equilibrium Thermodynamics, 21*(4), 307–320.

Zotin, A. A., & Zotin, A. I. (1999). *Direction, rate and mechanism of progressive evolution.* Nauka.

Zotin, A. I., & Zotina, R. S. (1993). *Phenomenological theory of development, growth and organism aging.* Nauka.

Zotin, A. A., Zotin, A. I., & Lamprecht, I. (2001). Bioenergetic progress and heat barriers. *Journal of Non-Equilibrium Thermodynamics, 26*(2), 191–202.

Chapter 4
Howard Odum and the Evolution of the Maximum Power Principle

Abstract This chapter explores H. T. Odum's development of the maximum power principle (MPP), including its relation to systems ecology, speed and efficiency, feedback loops, the pulsing paradigm, the quality of energy, and the maximum entropy production principle. This chapter reviews the evidence that supports the MPP and considers criticisms of it. Finally, this chapter discusses Odum's use of the MPP to critically analyze moral and religious values.

4.1 Introduction

While H. T. Odum (1924–2002) describes Alfred Lotka as the source for the Maximum Power Principle (MPP), he developed it further in a number of substantive ways. He applied it to all kinds of natural systems and ended up doing more with it than any other scientist. He is also the scientist that eventually began to call the principle the *maximum power principle*.

Odum's father, Howard W. Odum, was a broad-thinking, empirically oriented and quite famous American sociologist at the University of North Carolina. He encouraged H. T. and his brother Eugene to go into science and develop new approaches that would further social progress. Both brothers went on to become extremely influential ecologists. After serving in World War II, H. T. earned a Ph.D. in zoology at Yale University, working under the supervision of the British ecologist G. Evelyn Hutchinson, who was sometimes referred to as the "father of modern ecology." Hutchinson was a broad ranging ecologist and his approach had a large impact on H. T. Odum. H. T. was extremely prolific. He published 12 books and over 350 scientific papers (O'Neill, 1996), including a number of papers that played a significant role in the development of areas of research that are now distinct fields: ecological modeling (Odum, 1960); ecological engineering (Odum et al., 1963); ecological economics (Odum, 1971); estuarine ecology (Odum & Hoskin, 1958); tropical ecosystems ecology (Odum & Pigeon, 1970); and general systems theory. The ecologists Mark Brown and David Tilley have done a series of interviews with ecologists that worked with Odum as Ph. D. students or

© The Author(s), under exclusive license to Springer Nature
Switzerland AG 2025

T. McWhirter, *Maximum Power and its Philosophical Roots*, SpringerBriefs in Energy, https://doi.org/10.1007/978-3-031-80622-3_4

post-doctoral researchers which are available on Youtube under the title—*The Legacy of H. T. Odum and System Ecology*.[1] One of the things commonly mentioned in these interviews is how Odum opened their mind to new ways of thinking about science; he was developing a theory of how the world works and a central feature of it was energy and the MPP.

This chapter will review H. T. Odum's development and use of the MPP. It will consider how the MPP relates to Odum's systems approach to ecology and his energy systems language, review the work Odum did with the physicist Richard Pinkerton that described how the MPP involves a tradeoff between speed and efficiency, discuss some of the important differences between the MPP and Lotka's principle of maximum energy flux, and illustrate the relation Odum described between the MPP and feedback loops, the pulsing paradigm, and the quality of energy; it will describe the relation between the MPP and the maximum entropy production principle, review the empirical and theoretical evidence we now have that supports the MPP, critically analyze some of the criticisms of the MPP; and, finally, it will discuss Odum's view of how the MPP relates to the evolution of moral and religious values.

4.2   The MPP and Systems Ecology

There are a number of reasons why the MPP would be of particular interest to a systems ecologist. This principle relates to at least five different aspects of systems ecology: (1) the focus on synthesis rather than analysis; (2) the interdisciplinary nature of systems research; (3) the use of feedback loops; (4) the focus on energetics; and (5) the inclusion of human social systems.

First, one of the fundamental tenets of systems theory is that rather than analyzing one part of a system on its own and ignoring the other parts, the focus is on synthesizing how all the parts of the system work together in a way that is influenced by its environment: the focus is on the big picture. Lotka, for example, used a systems approach to explain how the principle of maximum energy flux (PMEF) can work together with James Johnstone's principle that in living processes the increase of entropy is retarded. He suggested that some parts of an ecological system, such as a plant or community of plants, could focus on capturing and preserving energy and other parts, such as animals, could focus on dissipating energy. All these parts can work together in a way that is consistent with the PMEF. The MPP is a further developed version of the PMEF and it operates in a similar fashion. It is not only the product of a thorough analysis of one part of a system; it emerges from an understanding of how different organisms and natural systems compete through evolutionary processes in different environments.

---

[1] *The Legacy of H. T. Odum and System Ecology* available at: https://www.youtube.com/playlist?li st=PLNc5B1CAPzA52yoxiIw72g9JF_70ykeTs

Second, systems ecology is fundamentally interdisciplinary: It includes physics, biology, and economics, among other disciplines. This approach enables scientists to develop a more comprehensive view of the way systems operate. The first and probably the most important paper Odum wrote on the MPP was "Time's Speed Regulator: The Optimum Efficiency for Maximum Power Output in Physical and Biological Systems," which was published in the *American Scientist* in 1955. Odum co-wrote this paper with the physicist Richard C. Pinkerton. This partnership helped Odum explain how the MPP applied to physical systems as well as biological and ecological systems. Even in the work Odum did on his own, his research was exhaustive and, in many cases, uniquely interdisciplinary. He later said that one of the most important and unfortunate issues that affected his work was the premature death of Pinkerton.

Third, feedback loops provide a mechanism that explains nonlinear changes in the evolution of natural systems from a systems perspective. They help explain how the energy cycle of systems evolves over time in a pulsing paradigm. When the power of natural systems is maximized through evolutionary processes, as outlined by the MPP, it will often be reflected in nonlinear changes in the energy cycling through systems. By explaining how these changes occur, feedback loops help explain how the MPP operates. Feedback loops and their relation to the MPP will be discussed in more detail later in this chapter.

Fourth, systems ecology focuses on developing an understanding of how energy moves through systems. The PMEF and the MPP describe how the energy cycling through systems evolves over time. Given this focus on energetics, one can see that when systems ecology begins to emerge in the twentieth century, in the wake of the rise of evolutionary theory in the nineteenth century, it would eventually lead to what has been called the "thermodynamic school" of evolution and the PMEF and the MPP (Fry, 1995, p. 229).

Fifth, systems ecology includes human social systems in its study of ecological systems. This has only become more important as scientists have begun to better understand the impact contemporary human societies are having on ecological systems around the world (see, e.g., Steffen et al., 2015). Lotka describes the PMEF applying to human social systems and Odum describes the MPP in the same way.

Odum used "energy network diagrams" as a language to describe "energy flows and forces" that make up the "parts and relations of all systems" (1971, p. 37). He used a whole list of symbols to represent different aspects of systems: an energy source, connecting lines for energy flows, and unique symbols for passive storage, heat sink, potential generating work, cycling receptors, work gate, self-maintenance, economic exchanges, and so on (1971, p. 38). With a little practice, these diagrams made it easier to see and understand how energy moves through systems. They provide a kind of universal language of energy systems (Cevolatti & Maud, 2004). Odum used these diagrams to, among other things, illustrate the feedback loops that enable systems to maximize power. We can see Odum's development and use of this energy network language as a byproduct of his systems approach to ecology: this language makes it easier for scientists from different disciplines and countries to better understand how the different parts of systems work together.

4.3   The Efficiency and Speed of Maximum Power

In the 1950s, Odum began to wonder why the efficiency of photosynthesis was quite low and his efforts to answer this question led him to work with the physicist Richard Pinkerton. The paper they published in 1955 further develops our understanding of the MPP in a fundamental way by exploring in further detail aspects of Lotka's PMEF and by adding a few additional twists. In the addendum to his essay "Contributions to the Energetics of Evolution," Lotka described how the selective advantage of efficiency increases when the amount of available energy is limited; when the available energy is not limited, the energy flux can be maximized through processes that are less efficient. Odum and Pinkerton explore in greater detail the quantitative relation of efficiency and speed in evolutionary development.

Odum and Pinkerton describe Lotka proposing a "'law of maximum energy' for biological systems" (1955, p. 332). Based on this law, they propose a postulate:

> Under the appropriate conditions, maximum power output is the criterion for the survival of many kinds of systems, both living and non-living. In other words, we are taking "survival of the fittest" to mean persistence of those forms which can command the greatest useful energy per unit time (useful power output). (1955, p. 332)

In this way, Odum and Pinkerton further develop Boltzmann's and Lotka's reformulation of Darwin's theory of natural selection. Their investigation seeks to demonstrate the efficiency and speed (or rate) of energy transformations that produce the maximum useful power output.

To illustrate the results of their research in a manner that is easy to understand, Odum and Pinkerton begin by focusing on 'Atwood's machine:' two baskets that are attached by a rope that runs over a pulley. This machine was often used in introductory physics courses at the time. When a heavier weight is placed in one basket, the machine can be used to move a lighter weight to the top of the system. The system can be used to explore how different ratios of the weights affect the useful work that can be done over time.

One can imagine using a system like this to move coal or gold from the bottom of a mine to the surface (Hall, 2017, p. 137–138). If the weights in the baskets are nearly the same, the load the machine is attempting to lift moves very slowly; the useful power output approaches zero and the efficiency approaches 100%. On the other extreme, as the weight of the load the machine is attempting to move approaches zero, the load in the heavier basket will fall at a maximum rate, but it will move hardly anything, so the efficiency of the process also will approach 0% and the useful power output will also approach 0%. Much of the input energy will be expended in the form of heating the substrate as the descending load crashes into the ground. In between these two extremes, there is an optimum ratio between the weights of the two baskets which moves a moderately heavy weight at a moderately high velocity and does the most useful work. Odum and Pinkerton note that one can use 'Atwood's machine' in a physics laboratory to demonstrate that the rate of efficiency that maximizes useful power output (i.e. moving the load uphill) in this ideal case is 50% (1955, p. 333; fig. 2). Subsequent investigations by other scientists have suggested that Atwood's machine maximizes useful power output at 61.8% efficiency (Smith, 1976) and under certain environmental conditions the maximum useful power output of energy transformation is reached at levels of efficiency that are as high as 67% (Silvert, 1982).

We see this general idea of a tradeoff between speed and efficiency throughout our everyday life. Odum provides a common example: if a truck is overloaded, it will hardly be able to move; if there is nothing in the truck, it will go fast but it will not deliver very much; the maximum amount of useful power (product delivered per time) is created when there is an intermediate level of load and speed (2001, Time's Speed Regulator). Similarly, when riding a bicycle, if you set the gear too low, each turn of the pedals will be efficient but it will be difficult to make; if you set the gear too high, it will be easy to turn the pedals, but each turn will not move you very far. In between these two extremes, there is a gear that enables you to do the most useful work (Hall, 2017, p. 143). Odum and Pinkerton use a number of diagrams to illustrate how this same thesis applies to many other processes: a water wheel turning a grindstone; one battery charging another battery; a thermocouple running an electric motor; a thermal diffusion engine; the metabolism of a pseudo-organism (with no self-repair); food capture by an organism for its maintenance; a model of photosynthesis; primary production in a self-sustaining climax community; and the growth and maintenance of a civilization.

Odum explains in another publication how this thesis applies to photosynthesis, enabling him to answer the question he initially sought to answer at the outset of his research: why is the efficiency of photosynthesis so low? During photosynthesis, the energy associated with incoming photons from the sun "separate electrons from organic structure, which gains a plus charge as a result" (2007, p. 37). "The plus charges of most plants interact with water to make oxygen." The negative charges "drive the load of organic production biochemistry." In laboratory experiments, isolated chloroplasts were analyzed and their different reaction loads were measured. The results demonstrated that the efficiency of photosynthesis varies in a manner that is inversely proportional to the intensity of light, and Odum suggests this applies to large ecosystems as well (2007, p. 40, Fig. 3.3(c)).

Plants have the ability to add or decrease green chlorophyll in order to alter the efficiency of photosynthesis. In environments where the intensity of light is high, plants will reduce chlorophyll, reducing the efficiency of photosynthesis and the plants there will be a lighter shade of green. In those environments where the intensity of light is low, such as Oregon in the winter, the plants will add more Chlorophyll to increase the efficiency of photosynthesis and they will, as a result, be a darker shade of green. Odum once stated that "the reason Ireland was so green was that the cloudy weather forced the chlorophyll to be near the surface of the leaves."[2] By adjusting the efficiency of photosynthesis in this manner, plants are able to adapt to the conditions in their environment and reach that optimum level of efficiency that maximizes the output of useful power, given the energy that is available. The reason

---

[2] This is from a conversation with Charles Hall.

photosynthesis is not as high as it could be is because in the evolutionary process, there is selection for the optimum rate of efficiency for the maximum output of useful power, and in many environments, there is an abundance of sunlight and therefore this optimum rate of efficiency is quite low.

Hall notes that the competitive demands of evolutionary environments can place a high value on speed rather than efficiency; if an organism takes too much time to exploit a source of energy, a competitor that is more rapid will have a selective advantage (2017, p. 143). In many environments, speed will be more important than efficiency for maximizing useful power output. However, the biologist Doug Glazier argues that speed can have negative consequences: it can increase the possibility of injury, bad decisions, and exposure to predators (2024, p. 33, 35). So, the optimum rate of speed for maximizing useful power output can vary in different environments.

At the end of their essay, Odum and Pinkerton provide an example of how the optimum level of efficiency for maximizing useful power output can vary in different environments. They write, "under conditions of limited raw materials as found in many areas of the world, a higher efficiency is the best arrangement…" for human social systems (1955, p. 343). They also acknowledge that in environments where the intensity of light is low, the efficiency of photosynthesis can get as high as 90% (1955, p. 341). If the available materials and energy are limited, useful power output is maximized by using more efficient methods. Odum makes this point throughout his career (Odum & Pinkerton, 1955, p. 341, 343; Odum, 1982, p. 35; Odum, 2001, 2007, p. 215). Near the end of his life, he wrote a book entitled *A Prosperous Way Down* (2001) in which he considers the changes that will be necessary to maximize power when our ability to use fossil fuels begins to diminish. He writes that when the conditions enable growth, then all of the things associated with growth—such as competition and exploitation—are considered good; when growth is not possible and descent is necessary, then all the things associated with efficiency come to be regarded as good (2001, Public Perception). Odum does not describe this position as an exception to the MPP; it is described as a way power is maximized when the available energy is limited (1982, p. 35).

Odum's position reflects an understanding of an important difference between efficiency and speed. Increasing the useful power output by increasing the speed (or rate) of energy transformations uses more of the available energy; increasing the useful power output by increasing efficiency does not require using more of the available energy. This difference is not a significant issue when the available energy is abundant, but when the available energy is low, the ability to increase useful power output through increases in rate are constrained; the ability to increase useful power output by increasing efficiency is not constrained in the same way. This is why the optimum efficiency for maximizing useful power output increases as the available energy decreases: this approach utilizes "the maximum power available" (Odum, 1982, p. 35). Contemporary scientists who have a unique familiarity with Odum's work such as Mark Brown, who worked with Odum as a Ph.D. student and closely thereafter, understand this aspect of the MPP (Brown, 2023); however, as we will see in Sect. 4.12.2, many contemporary scientists overlook this aspect of the MPP.

The relation Odum and Pinkerton describe between available energy and the optimum level of efficiency needed to maximize useful power output qualifies their conclusion that the optimum level of efficiency will "never exceed 50 per cent of the *ideal* "reversible" efficiency" (1955, p. 332; my emphasis). Their argument is that useful power output can be maximized the most under conditions of intermediate light intensity and intermediate efficiency and, in this *ideal* case, the efficiency will never exceed 50%; however, in order for plants to maximize useful power output in environments where the intensity of light is far below the optimum intermediate level, they must increase the efficiency of photosynthesis, in some cases to levels well above 50%, and this fact applies to all natural systems. This illustrates Lotka's point that the fact that PMEF applies to systems in a manner that is "compatible with the constraints to which the system is subject" "modifies" the principle in a manner that is worthy of "further consideration" (1922, p. 148, 150). This point also applies to the MPP and we will give it further consideration later in this chapter (particularly Sect. 4.12.2).

4.4 Maximum Useful Power Output

4.4.1 Power Output

While Odum and Pinkerton follow Lotka's lead in a number of respects, there are important differences in their approach. First, Odum and Pinkerton argue that there is selection for those natural systems and organisms that maximize useful power *output*. Lotka, on the other hand, describes the PMEF in terms of the flow of the energy through systems including: the "available energy absorbed by and dissipated within the system per unit of time" (1922, p. 150, n. 5). However, in that same essay, Lotka also writes that in the struggle for existence, "the advantage must go to those organisms whose energy-capturing devices are most efficient in directing available energy into channels favorable to the preservation of the species" (1922, p. 147), and, as mentioned, he also writes that it now appears probable that human beings have "been unconsciously fulfilling a law of nature according to which certain quantities tend toward a maximum" and that quantity is "power, or energy per unit of time" (1921, p. 172, 1922, p. 149). These statements appear to suggest that power *output* provides a selective advantage. Furthermore, as mentioned, Lotka suggests that where the supply of available energy is limited, the selective "advantage" will go to organisms that are more efficient. When a natural system increases its efficiency, it increases the output of power, but it does not necessarily increase the energy flux through systems because it does not necessarily increase the amount of energy absorbed or captured by the system. Here, once again, Lotka appears to be implying that organisms that increase power output have a selective advantage. So, the picture Lotka provides is not completely clear.

We can actually see Odum's (and Pinkerton's) shift to maximum power output as being a natural development of the PMEF when it is used to consider the value of efficiency. Lotka ends his 1922 paper describing the value of efficiency. Odum's efforts to understand why the efficiency of photosynthesis was so low led him to work with Pinkerton. In their 1955 paper, they describe the relation between efficiency, speed, and the MPP. If the focus is on the amount of energy cycling through systems, the energy flux, the full value of the efficiency cannot be measured directly. The impact of efficiency can be measured directly, however, by focusing on useful power output.

This difference between the PMEF and the MPP can be mitigated to a degree when we take into consideration the fact that much of the energy dissipated by natural systems and organisms is used to acquire additional energy. When the energy used for this purpose is expended in a more powerful manner, it can increase the energy that moves through natural systems and organisms. Odum and Pinkerton describe all energy transfers as a "combination of two parts": an input and an output (1955, p. 332). The input can affect the output and vice versa. When we acknowledge that natural systems and organisms regularly expend energy in order to acquire energy, we can see that the PMEF and the MPP are very closely related in spite of this difference in the way they are defined.

4.4.2   Useful Power

Another important difference between the PMEF and the MPP is that the MPP is described by Odum and Pinkerton as maximizing the output of *useful* power. Lotka describes the PMEF maximizing the energy moving through systems, and he does not distinguish between the energy dissipated simply due to the second law and the energy dissipated in a useful manner. Odum's and Pinkerton's analysis of Atwood's machine makes this distinction. When the weight of the load in the basket to be lifted is reduced to zero, the load in the other basket falls to the bottom with great speed; it does not lift anything, and much of the energy is lost in the form of heat. The heat that is lost in the process is not considered to be useful work; the work that is done to lift a load is considered useful work. This distinction between useful work and work that is not useful is not found in Lotka's description of the PMEF.

4.5   Feedback Loops

Feedback loops are parts of systems in which some portion of the system's output has an effect on the system's input, either positive or negative. A simple example would be what happens when glaciers in the oceans melt. Glaciers on the ocean reflect the sun's light and reduce the temperature at sea level. When glaciers melt, they are no longer able to reflect the sun's light, so absorption is increased and the temperature at sea level increases. This increase in temperature increases the tendency of other glaciers in the surrounding region to melt. When glaciers begin to melt, it creates a positive feedback loop that enhances the tendency of glaciers to melt. These kinds of feedback loops are often used to explain the non-linear changes in the evolution of complex systems. Likewise, useful energy absorbed by a plant can be used to make more leaves (or roots) that can increase that tree's ability to capture more energy.

Odum describes feedback loops as being fundamental to the MPP. Natural systems maximize their power by creating feedback loops that enhance their ability to take in and use energy. One simple example Odum uses would be a town that has a dam and a power plant. The town uses the power from the dam and the power plant to produce machinery and services that improve the operation of the dam and the power plant, enabling them to produce more energy for the town (1976, p. 42). In this example, the system takes in energy and performs operations that enhance the quality of this energy. Some of this higher quality energy is then used to enhance the system's ability to take in more energy: This process is called *loop reinforcement*.

Lotka provides a description of a feedback loop that is fundamental to the development of human civilizations, although he does not use the term feedback loop. He describes human beings in the past developing "ingenious contrivances" that greatly enhanced their ability to take in and use energy (1945, p. 188). The excess energy and time afforded by these "ingenious contrivances" left people with "a large balance available for 'play' activities and luxuries." Scientific research would have to be classified as one of these luxuries, which was initially done primarily out of curiosity. But this research, in turn, resulted in "that complete recasting of methods of production which is known as the industrial and agricultural revolution." These revolutions have dramatically increased the power of contemporary human social systems and provided them with an even larger balance of excess energy and time for "'play' activities and luxuries," such as scientific research. Odum describes feedback loops like this as enabling natural systems to maximize their useful power output. A contemporary example is geological science development leading to the exploitation of far more petroleum.

4.6   The Pulsing Paradigm

Odum recognized early on that natural systems do not develop in a linear fashion; their development is often structured by a pulsing rhythm. The seasons provide the simplest illustration of this rhythm. Agriculture is fundamentally structured by this cyclical rhythm of seasons, and part of the reason that temperate agriculture tends to be more productive than tropical agriculture is that people can slip in their cultivars while competing plants are dormant. Our life is structured by the rhythm of day and night. Many forested ecosystems grow, mature, and senesce, or burn, over cycles of centuries. There is a pulsing rhythm in chemical binding that occurs at the molecular level and the variations of carbon concentrations in the atmosphere over millions of years that dramatically transform the environments on the planet (Odum, 1983a, p. 574). Odum describes the pulsing rhythm as a fundamental part of nature at all scales, both spatial and temporal.

Although the geologist Gene Hunt does not mention H. T. Odum, he does use the term "pulsed" to refer to the rhythm of change outlined in the theory of punctuated equilibrium (Hunt, 2008). The fossil record appears to demonstrate there is a pulsing rhythm to evolution itself: there are long periods of equilibrium punctuated by short periods of dramatic evolutionary change (Eldredge & Gould, 1972, 1993). The idea of pulsing systems has also been developed by other scientists. Ilya Prigogine and Isabelle Stengers describe the tempo of punctuated equilibrium as being fundamental to the development of all natural systems (Prigogine & Stengers, 1984, p. 169–170). The philosopher of science Thomas Kuhn describes this same rhythm in the development of scientific knowledge: scientific paradigms guide research for long periods of time in what Kuhn calls "normal science"; periodically, these paradigms are replaced during scientific revolutions in what he calls "revolutionary science" (Kuhn, 1962, p. 5–10). Other scientists have also discovered the rhythm of punctuated equilibrium in the evolution of language (Atkinson et al., 2008). Hunt's use of the term "pulsed" points toward the structural similarities between the pulsing rhythm described by Odum and the theory of punctuated equilibrium.

Odum and J. R. Richardson (1981) argue that when there is an intermediate level of available energy, the pulsing rhythm maximizes power. In 1995, Odum published research he did with his brother Eugene on the pulsing paradigm in tidal salt marshes, tidal freshwater marshes, and seasonally flooded fresh-water wetlands (Odum et al., 1995). In his later work, Odum describes the principle of selection being guided by the maximum empower principle with pulsing (2007, p. 270). Periodic floods or fires are perceived as disasters to humans, but they are fundamental part of nature.

4.7 Maximum Empower

It is not clear exactly when Odum first started thinking about the quality of energy, but it seemed to be long incubating. He did research in the 1950s and 1960s that could have stimulated his interest in the subject. In *Environment, Power and Society*, he described coal and oil as "concentrated inputs of power" that were over the years the products of "billions of acres of solar energy" (1971). He was one of the first to point out that while solar energy was abundant, it was available in a form that was too dilute and the energy required for concentration was extremely high. Odum was familiar with the concept of "net production" and how it suggested that the energy provided by a source must exceed the amount of energy used to acquire it in order to provide a net benefit to society. This led to the concept of "net energy" which held that the actual value of energy to society is "the net energy, which is that after the costs of getting and concentrating that energy are subtracted" (Odum, 1973, p. 220). In the middle of the 1980s, a Ph.D. student that worked with Odum, Charles A. S. Hall, along with C. J. Cleveland and R. Kaufmann, defined the concept of the Energy Return on Investment (EROI) (Hall et al., 1986): it provides a ratio of the energy provided by an energy resource and the energy invested in order to get it.

Odum began to see forms of energy in terms of a hierarchy: the least useful forms, such as sunlight, which is dilute at the bottom; the more useful forms of energy, such as fossil fuels at the top. He believed the quality of a form of energy could be measured by the amount of energy required to develop that form of energy and he developed the idea that a system would pay for the energy loss (i.e., the process would be selected) only if its quality were proportional to that energy lost (1977b). Starting in the early 1980s, he used the concept of "embodied energy" to refer to the amount of energy used to develop a form of energy. Under the influence of David M. Scienceman, Odum eventually began to use the term *"emergy"* to refer to the amount of energy used to develop a heat unit (Kcal or Joule) of a product or service. He referred to the flow rate of *emergy* as *"empower"* and he further developed his MPP in order to take into consideration the quality of energy: he referred to the new principle as the *maximum empower principle* (MEP). Odum believed that the MEP and the MPP would coincide because the use of higher quality sources of energy and reinforcing energy cycles would increase the flux of energy through systems and the efficiency of their ability to transform lower-quality sources of energy (Odum, 1995, p. 318; Cai et al., 2004, p. 117). For example, a tree colonizing an old field might develop a higher quality fruit that would enhance a higher quality bird to the site. The bird would then seed in the new region in the field as it ate the fruits (and defecated other tree seeds) so that in time an ecosystem that could capture more of the sun's energy would develop.

As energy is used to increase the quality of a form of energy, the quantity of the form of energy is decreased. Odum provides some energy quality evaluations using the scale of equivalent costs in units of solar energy to put human services into perspective—illustrating how much energy is required to produce these forms of energy "in calories of solar equivalents per calorie" of the form energy in question: plants require 1000 calories of solar equivalent calories per calorie of plant energy; "coal, 2000; electricity, 7200; human food, 20,000; simple human service, 200,000; educated human service, 2,000,000" (1977a, p. 121). The highest quality forms of energy are kinds of information that require a great deal of energy to produce but themselves contain only a minute amount of energy.

This new focus on the quality of energy and the concept of emergy was, at the time, and is still, controversial. Not all scientists agreed with the approach and even some of his own students have resisted this aspect of Odum's work. But many scientists believe this new avenue is extremely important to the future of energy analysis and research. There is an Emergy Society that has Emergy Synthesis Research Conferences at the University of Florida, where Odum worked for many years. The society and the conferences are organized by Odum's former Ph.D. student, Mark Brown. The conferences started in 1999 and continue to this day, with many very interesting papers.

4.8   Definitions of the MPP

Over the course of his long career, Odum offered a number of different definitions for the MPP, which, in some cases, illustrate how the concept has evolved over time. In 1955, Odum and Pinkerton described the MPP as a postulate they made that is based on Lotka's proposed "law of maximum energy" for biological systems. They defined this postulate as follows:

> Under the appropriate conditions, maximum power output is the criterion for the survival of many kinds of systems, both living and non-living. In other words, we are taking "survival of the fittest" to mean persistence of those forms which can command the greatest useful energy per unit time (power output). (1955, p. 332)

In 1971, Odum described the MPP as a "general energy law" Lotka developed that holds that the "maximization of power for useful purposes was the criterion for natural selection" (1971, p. 32). In 1977, Odum described the MPP as holding that "the more lasting and hence more probable dynamic patterns of energy flow or power (including the patterns of living systems and civilizations) tend to transform and restore the greatest amount of potential energy at the fastest possible rate" (1977a, p. 109). In 1983, Odum described the MPP as a principle developed by Lotka that suggests that "systems prevail that develop designs that maximize the flow of useful energy" (1983a, p. 6). This definition implies that Lotka had defined the PMEF in terms of *useful* energy. However, this is not the case: Lotka defined it in terms of *available* energy.

In 1995, Odum wrote that Lotka "suggested that the MPP was a fourth law of thermodynamics" (1995, p. 311; cp. Fleck & Morel, 2006). To be more precise, as mentioned in the last chapter, Lotka had suggested that the principle of natural selection functions like a fourth law of thermodynamics and he believed it was guided by the PMEF. In the same year, Odum goes on to write that the MPP can be stated as follows: "during self-organization, system designs develop and prevail that maximize power intake, energy transformation, and those uses that reinforce production and efficiency" (1995, p. 311). Once again, this definition comes very close to Lotka's definition of the PMEF. In 2001, Odum described Lotka suggesting the "maximum power concept as a fundamental energy law rephrased here as follows: 'By trial and error many alternatives start to function, but only those designs that contribute more useful energy flow get reinforced and thus selected to continue'" (2001, 1361). The use of the term "useful" here distinguishes this definition from that provided by Lotka. In 2007, Odum writes, "systems that prevail are those with loading adjusted to operate at the peak of the power efficiency curve .... During self-organization, these systems reinforce (choose) pathways with optimum load for maximum output" (2007, p. 37). This definition appears to be more in line with that outlined by Odum and Pinkerton in 1955. In a footnote (2007, p. 37, n. 3), Odum adds, "because every real process requires power, the maximum and most economical collection, transmission, and use of power must be one of the primary selective criteria." In 2007, Odum also described the MEP as an amended version of Lotka's principle which holds that "*self-organization develops designs to maximize empower of each scale at the same time*" (2007, p. 89). Some of these definitions emphasize different aspects of the MPP; some restate the principle in different ways. Taken collectively, they provide a comprehensive picture of what Odum was thinking about when he refers to the MPP.

4.9   The Maximum Entropy Production Principle

In the introduction to this book, I discussed how, in the general scientific community in the 1970s, the maximum entropy production principle (MEPP) began to be used rather than the MPP (Paltridge, 1975). This continued over the following decades (Ulanowicz & Hannon, 1987; Swenson, 1989) and after the turn of the century, Roderick Dewar provided a theoretical basis for the MEPP and derived a provisional proof of it for both biotic and abiotic systems (2003, 2005). The principle came to be used by scientists to describe phenomena in several different scientific disciplines (Martyushev & Seleznev, 2006).

Maximizing useful power and maximizing entropy production are two different things, but they are fundamentally related. Power is a measure of the expenditure of energy over time; when energy is expended, work is done; when work is done, entropy is produced. Consequently, when the power of systems is increased, their ability to production of entropy is also increased. Lotka and Odum were aware of this. Odum writes that maximum entropy production "is another way of referring to maximum power utilization if feedbacks couple the products of power use to power generation" (1983b, p. 75). The MPP and the MEPP do focus on different aspects of the function of systems. The MEPP's focus on entropy may help people understand how natural systems can develop in a way that is consistent with the second law of thermodynamics. Some scientists have suggested that MEPP can be viewed as a "corollary" of the second law (Martyushev & Seleznev, 2006, p. 3).

Later in this chapter (Sect. 4.12.3), we will briefly consider the relative value of the MPP and the MEPP. What we can say, at this point, is that scientists that use the MEPP acknowledge its historical lineage and they give credit to Lotka and Odum (Martyushev & Seleznev, 2006, section 3.4.2; Kleidon & Lorenz, 2005, p. 15). Lotka developed the PMEF; Odum and Pinkerton developed the MPP; Dewar and other contemporary scientists have developed the MEPP: our understanding of the evolution of natural systems continues to evolve and we have no reason to think this will change.

4.10   The Evidence

Enrico Sciubba writes that no one has published any proof of the applicability of the MPP (2011, p. 1347, 1351). Charles Hall, who worked with H. T. Odum as a Ph.D. student, writes that Odum did not really do empirical studies that tested the

MPP; he even admitted that "it was extremely difficult if not impossible to test it directly as is the case with Darwinian selection more generally" (Hall, 2004, p. 110). Odum does, however, cite the work on streams by Sugita (1951) and Leopold and Langbein (1962) as providing some support for the MPP. Odum writes that Sugita "uses examples such as stream capture in land drainage" to suggest that "systems tend to maximize power in their organizational work" (Odum, 1983a, p. 118). Odum writes that Leopold and Langbein describe how the steepness and meander of streams develop in ways that help "maximize the power of a region's total energy flows" (Odum, 1983a, p. 118). Rather than dumping all the water straight down, releasing all of its energy at the end, streams tend to meander in a manner that is less steep, which gathers more energy and spreads it farther.

In 2006, Cai et al. published research they did in order to investigate empirically the MPP using plankton microcosms (Cai et al., 2006). They note that this principle holds that system designs will evolve that are able to capture a previously unused source of energy. They set up microsystems that developed under a fixed schedule of light duration and others where the light duration was changed based on water column pH. They found that the microsystems that developed under the pH controlled light were able to increase their light duration much more than the other microsystems; they were able to adapt in order to access an energy source that was not available to the other microsystems. Their research supported the MPP.

In 2013, Hall, Nancy Harris, and Ariel Lugo published the results of empirical surveys they did along an elevational gradient in the Luquillo forest in Puerto Rico. They used a LI-COR CO_2 analysis machine, rock climbing technology, and a giant slingshot to take measurements throughout the canopy at different elevations. The results demonstrated that there was a tradeoff between efficiency and rate over the elevational gradient and that maximum net photosynthesis was found at intermediate elevations as they had hypothesized based on the MPP (Hall et al., 2013).

In 2016, the scientists Timothy M. Lenton, Peter-Paul Pichler, and Helga Weisz published research they did on six energy revolutions, three in the history of the earth and three in human history: in earth history, they analyzed the origin of anoxygenic photosynthesis, oxygenic photosynthesis, and eukaryotic photosynthesis; in human history, they analyzed the Paleolithic use of fire, the Neolithic revolution, and the Industrial revolution (2016). They used an extraordinary array of empirical evidence and theoretical models to "quantify the resulting increase in energy input to the biosphere or to human societies" for each of these energy revolutions. They did not mention the PMEF, the MPP, or the MEPP, but their conclusions are consistent with them.

The Neolithic and Industrial revolutions provide evidence of the role of efficiency and speed (i.e., increasing capacity) in evolutionary development. The energy return on investment (EROI) of hunter gatherers has been estimated at around 10:1 (Hall, 2017, p. 162). The EROI for the entire English society during the agricultural period from 1300 to 1750 has been estimated at around 2.5:1 (Hall, 2017, p. 167). In spite of this reduction of EROI, from 1 AD to 1750 AD the global population doubled. This can be explained by the large increase in the total amount of exergy that could be captured and used through agriculture and the domestication of animals. Even though the EROI declined, power was increased by increasing capacity or speed; using Lotka's vernacular, they were able to "enlarge the wheel" to such an extent that it more than made up for the fact that the wheel was spinning slower.

The situation was different during the Industrial Revolution. The EROI of global fossil fuels from 1800 to 1920 was approximately 30–40:1 (Hall, 2017, p. 166; Court & Fizzaine, 2017). After 1920, the EROI of oil and coal went up to as much as 60:1. The discovery of fossil fuels presented human societies with a unique opportunity to utilize a form of exergy that enabled them to increase both the capacity (or speed) and the efficiency of our ability to use exergy: they could "enlarge the wheel" *and* make it "spin faster." While the value of efficiency goes up when exergy is limited, it still has a value when exergy is abundant: it can still maximize useful power output. The Neolithic and the Industrial revolutions demonstrate how both speed and efficiency can be used to maximize power under certain conditions as outlined by Odum and Pinkerton.

The biophysicist Aleksandr Zotin has illustrated that the emergence of exosomatic exergy dissipation in human evolution is consistent with the PMEF, the MPP, and the MEPP (Martyushev & Seleznev, 2006, 3.4.2). It enables human beings to increase the exergy flux through human social systems to an extent that is far beyond that allowed by the thermodynamic limits of human bodies. So, the emergence of exosomatic exergy dissipation in the evolution of human beings is itself evidence that supports the conclusion that this evolution is guided by the PMEF and the MPP.

If we consider the different ways human beings have expended exergy outside their bodies historically, we can see that there is evidence that supports extending the thesis developed by Lenton et al. (2016) to the first three industrial revolutions. Below is a partial, chronological list of the advancements in the power output of heat engine technology provided by Earl Cook (1976, p. 29).

Machine	Date	Horsepower
Man pushing a lever	3000 B.C.	0.05
Ox pulling a load	3000 B.C.	0.5
Vitruvian water mill	50 B.C.	3
Post windmill	1400 A.D.	8
Watt's steam engine (land)	1800 A.D.	40
Steam engine (marine)	1837 A.D.	750
Steam engine (marine)	1900 A.D.	8000
Steam engine (land)	1900 A.D.	12,000
Steam turbine	1906 A.D.	17,500
Coal-fired steam power plant	1973 A.D.	1,465,000
Nuclear power plant	1970 A.D.	1,520,000

This list illustrates that the power of these engines has continued to go up over time and there are non-linear increases that correspond to each of the first three industrial revolutions: first 1760–1840, second 1870–1914, third 1947–2009. The research by Lenton et al. and Cook provides evidence that ecological systems and human social systems have evolved in a manner that is consistent with the PMEF, the MPP, and the MEPP. Their research supports these principles in a historical fashion that is similar to the way the fossil record supports the concept of punctuated equilibrium (Eldredge & Gould, 1972, 1993).[3]

This extension of Lenton et al.'s thesis to the first three industrial revolutions is also supported by research published in 2015 by Will Steffen, the director of the Australian National University Climate Change Institute. He and his team of scientists and researchers uncovered a wealth of evidence which demonstrates that from 1750 to the present, there has been a dramatic increase in primary energy use, as well as a whole host of other socio-economic and earth systems trends: in the growth of the world population, the gross domestic production, foreign direct investment, water use, transportation, telecommunication, international tourism, atmospheric concentrations of carbon dioxide, nitrous oxide, methane, ocean acidification, tropical forest loss, and terrestrial biosphere degradation (Steffen et al., 2015). The socio-economic trends are listed below (Fig. 4.1). Steffen et al. refer to these trends collectively as the "Great Acceleration," and they argue that it defines the trajectory of the Anthropocene, the proposed geological epoch dating from the commencement of significant human impact on Earth's geology and ecosystems. As it turns out, Henry Adams' "*law of acceleration*" was just scratching the surface of this Great Acceleration.

In 2024, the scientist Elijah Thimsen published research he did on data from the Clouds and Earth's Radiant Energy System and Moderate Resolution Imaging Spectroradiometer instruments in order to determine how the rate of sunlight absorbed by the earth and the rate of global entropy production have evolved from 2002 to 2023 (Thimsen, 2024). The results demonstrate that the rates of sunlight absorption and entropy production both increased during this 20 year period. Thimsen argues that this evidence is consistent with both the MPP and the MEPP.

Over the past two decades, a number of other scientists have used the MEPP to explain phenomena in a number of various scientific disciplines, including ecology, chemistry, biology (the origin of life), sociology, and cosmology.[4] The scientists Leonid Martyushev and Vladimir Seleznev published a paper in 2006 entitled "Maximum Entropy Production Principle in Physics, Chemistry and Biology" (Martyushev & Seleznev, 2006). They suggest the discussions of the MEPP have been fragmented and, as a result, different research teams have been unaware of the studies performed by other scientists. One of the main objectives of their paper is to overcome this fragmentation. The scientists involved in this research have generally not focused on discussing the MPP; they have focused on applying the MEPP to the

---

[3] Recently a paper was published that used the multiscale thermodynamic perspective and the power laws associated with it to provide an explanation for the fractal structures found in nature in so many different scientific disciplines (Cordaro & Venegas-Aravena, 2024). This can be seen as further illustrating the explanatory power of the multiscale thermodynamic perspective and the power laws associated with it, like the MPP.

[4] For ecology, see Jorgenson et al. 2004 and Kleidon et al., 2010; for chemistry, see Fleck et al., 2006; for biology (origin of life), see Michaelian, 2011 and England, 2013; for sociology, see Ameniya et al., 2011; for cosmology, see Bonanno et al., 2011 and Farajollahi et al., 2011.

Socio-economic trends

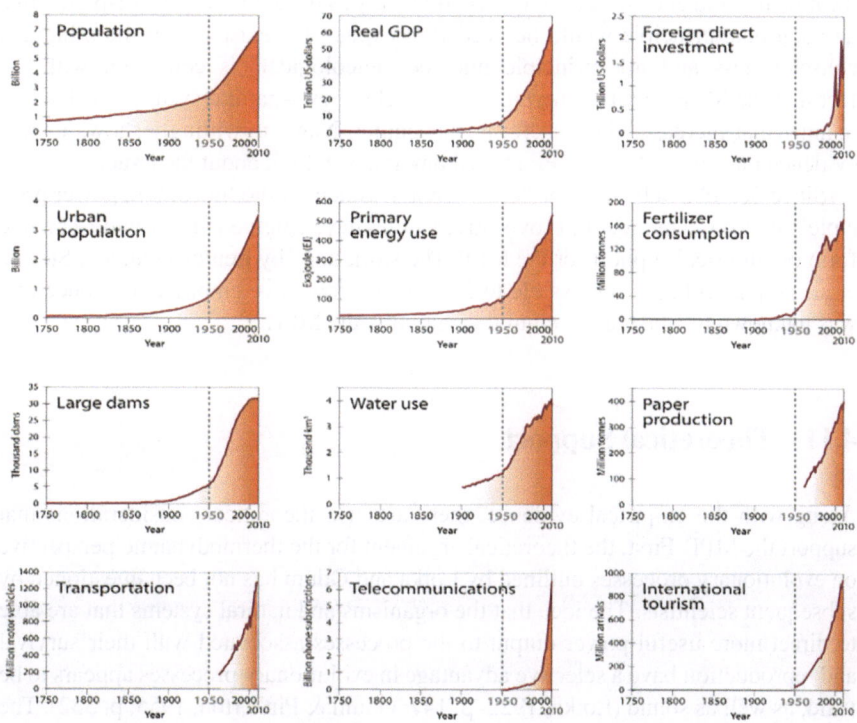

Fig. 4.1   Socio-economic trends (Steffen et al., 2015)

phenomena in their discipline. Since the MPP and the MEPP are fundamentally related, their research suggests that the phenomena in their disciplines could perhaps also be explained by the MPP.

The philosopher Thomas Nagel helps us recognize a different kind of evidence that supports the MPP. He argues in his book *Mind and Cosmos* (2012) that the contemporary approach to evolutionary theory, which characterizes it as a random process, is unable to account for the existence of the mind and consciousness and is therefore incomplete. Nagel's argument (which is controversial) is that we need to use different principles to account for the emergence of conscious life and they may be teleological. When we view evolution from a thermodynamic perspective that includes the MPP, we are in a better position to explain the existence of brains and consciousness. The brain and consciousness enabled human beings to envision and develop more and more powerful machines that have facilitated the nonlinear increase in the exosomatic dissipation of energy. This trend has enabled human beings to overcome the thermodynamic limitations of our bodies and increase our ability to dissipate energy to an unprecedented extent, relative to the other animals on this planet.

Nagel's argument would seem to be consistent with the MPP. The contemporary view of evolution as being a random process is not able to explain the existence of the human mind and consciousness as effectively as the MPP. Nagel's suspicion that an additional principle would be needed to explain the existence of the mind and consciousness and this principle may be "teleological" is consistent with the fact that the MPP entails a thermodynamic teleology—natural systems evolve in a manner that increases their useful power output. Nagel unwittingly shows us that evidence that supports the MPP is used anytime we think about the issue.

Since Sciubba published his "critical reassessment of the 'maximum power principle'" in 2011, a number of provocative studies have emerged that provide a unique form of empirical support for the MPP. The work done by Lenton et al. and Steffen et al., in particular, is impossible to ignore. We now have empirical evidence that was unknown to Lotka and Odum that supports the MPP.

4.11   Theoretical Support

Along with the empirical evidence, there are also theoretical considerations that support the MPP. First, the theoretical argument for the thermodynamic perspective on evolutionary processes outlined by Lotka and Odum has not been questioned by subsequent scientists. The idea that the organisms and natural systems that are able to direct more useful power output to the processes associated with their survival and reproduction have a selective advantage in evolutionary processes appears to be valid, as well as sound (Lotka, 1922, p. 147; Odum & Pinkerton, 1955, p. 332). The thermodynamic school of evolution has, indeed, grown since the beginning of the twentieth century.

Second, in retrospect, Henry Adams' application of Newton's first law of motion to cultural evolution now appears to be prescient, particularly when we consider it within the context of Lotka's description of the feedback loop in human social systems created by scientific research: as humans use scientific investigations to develop and use more powerful machines and methods to meet their needs, it gives humans more time and resources to do more scientific investigation to develop more powerful machines and methods. This feedback loop has propelled dramatic changes in human life over the past few centuries, as evidenced by the "Great Acceleration" described by Steffen and his colleagues. Unless there is another force that acts in opposition to this acceleration, we have no reason to think it will slow down. The fact that we will eventually reach the peak of our ability to use fossil fuels is a force that could oppose the momentum of this acceleration; in fact, many believe that this is highly likely. However, this merely confirms Adam's view and its relevance to the MPP: the cultural acceleration associated with the MPP, and the feedback loops associated with it, have a momentum that will continue to grow unless there is a force equally strong opposing it.

Third, Iris Fry, a professor at the Cohn Institute for History and Philosophy of Science and Ideas at Tel Aviv University, published a paper in 1995 entitled

"Evolution in Thermodynamic Perspective: A Historical and Philosophical Angle" and a book in 2000 entitled *Emergence of Life on Earth: A Historical and Scientific Overview*. In the paper, she argues that what she calls the "thermodynamic school" of evolution provides a "law-governed" view of the "continuous nature of evolution" that is better able to explain the origin of life on earth. She does not argue that scientists have succeeded in providing such an explanation; she argues that the scientists in the "thermodynamic school" have outlined an approach that appears to be more promising. Since she published her essay, other scientists in the "thermodynamic school" have made progress on the issue (see, for example, Michaelian, 2011). She refers to Lotka's research (1995, p. 228), but she focuses her attention on the work of Lawrence J. Henderson (1878–1942) and Jeffery S. Wicken (1942–2002). She does not refer to the MPP or the PMEF. However, these principles are products of this school in the sense that Lotka and Odum both view evolution from a thermodynamic perspective, and these principles contribute to the "law-governed" view of the "continuous nature of evolution;" therefore, they also put scientists in a better position to explain the origin of life.

4.12 The Critics of the MPP

4.12.1 The Problems with Empower

When Odum began to develop his view of the quality of energy and started further using the terms emergy and empower, it, among other things, increased the distance between his view and Lotka's, which was outlined in terms of available energy, or exergy. Contemporary scientists began to express some concerns with this turn in Odum's work. Sciubba argues that emergy and exergy analyses are not commensurable (2009; see also Herendeen, 2004). He also describes emergy analyses as facing difficult practical problems: how far back do we go in order to calculate all the energy that went into a product or a form of energy? There does not appear to be any clear answer to this question. Mark Brown acknowledges that some find it impossible to quantify the "amount of sunlight that is required to produce a quantity of oil" (Brown & Ulgiati, 2004, p. 210). So, the maximum empower principle has faced criticism.

One can acknowledge the practical problems with the concept of emergy and the maximum empower principle and still appreciate the importance of the distinction Odum is making in the quality of energy. The light from the sun is abundant, but it is dilute and cannot be used as easily as sugar or food or oil or electricity to do useful work. In the future, we may be able to overcome the practical challenges facing the concept of emergy. Elaborate computer models and artificial intelligence systems could be used to calculate the amount of sunlight that is required to produce a quantity of oil. These calculations may not be exact, but they could be close enough to recognize important distinctions that help us make better decisions now.

Furthermore, a scientist in the future could come up with a different way to calculate the quality of energy that is more feasible. The practical problems with the concept emergy need to be taken into account, but they do not provide a reason to deny the general importance of the distinction Odum makes in the quality of energy. Odum believed that in science the originators of an idea would lay it out and make initial estimates of parameters and that later practitioners would refine the early estimates.

4.12.2 *Reductive Interpretations of the MPP*

In contemporary critical investigations of the MPP, we find a number of reductive interpretations that overlook different aspects of the processes that maximize power. Sciubba ignores how under certain conditions, power can be maximized by increasing efficiency. At the end of his critical assessment of the MPP, Sciubba concludes that the weakness of Lotka's view of evolution appears to be his insistence on the validity of a maximum principle. Sciubba suggests that contemporary scientists are more inclined to believe that two principles are needed to explain the development of natural systems: systems near equilibrium tend to evolve with maximum efficiency, while those far from equilibrium tend to dissipate energy the fastest way possible (Sciubba, 2011, p. 1352). Sciubba's argument also applies to Odum's MPP.

Both of the principles that Sciubba describes contemporary scientists adopting can be considered consistent with the relation between power, efficiency, speed, and exergy that Odum and Pinkerton describe in their formulation of the MPP (1955); they are also consistent with Lotka's formulation of the PMEF (1922). As discussed earlier (Sect. 4.3), Odum has stated that when systems approach equilibrium and the available energy decreases, the optimum efficiency needed to maximize useful power output increases. For this reason, Odum argues that the minimum entropy production principle developed by Prigogine and Wiame is not an exception to the MPP; it is consistent with it, a special case of it: when the exergy is limited, systems that are more efficient and minimize entropy production will maximize the power that is "available" under these conditions (1982, p. 35, 1983a, p. 7, 1983b, p. 79). In his critical assessment of the MPP, Sciubba does not realize that the two principles he describes contemporary scientists adopting are both consistent with the MPP when it is understood as applying in a manner that is "compatible with the constraints" on systems (2010, p. 1352).

We find a related problem in a recent paper published by the biologist Douglas Glazier (2024). To be fair, he does not focus exclusively on the MPP developed by Odum; he discusses the "maximum power theory" we also find in the work of some contemporary scientists, such as G. J. Vermeij, J. H. Brown, and C. F. Jordan (2024, p. 35). He provides an impressive collection of the empirical evidence available on the relation between power and efficiency in living systems, which demonstrates that in some systems power and efficiency vary in a manner that is negatively correlated (or inversely proportional), and in other systems they vary in a manner that is positively correlated (they increase or decrease together). He uses this evidence to argue that the "maximum power theory" must be replaced because it assumes that power and efficiency always vary in a manner that is negatively correlated (2024, p. 2, 35). He argues that the MPP is really a "maximum power at intermediate efficiency principle" because it describes power being maximized at the expense of efficiency (2024, p. 3).

Glazier acknowledges that Lotka recognized that when resources are limited, natural selection will favor organisms and natural systems that are more efficient, and he acknowledges that Lotka described the PMEF applying to natural systems in a manner that is "compatible with the constraints" on them (2024, p. 17, 13); however, Glazier goes on to say that Lotka "never fully explained what these constraints are." This last point Glazier makes is correct, but it has no bearing on Lotka's definition of the PMEF: Lotka does not say that this principle applies to natural systems "so far as compatible with the constraints" on systems that he has "fully explained;" he writes, "*natural selection* tends to make the energy flux through the system a maximum, so far as compatible with the constraints to which the system is subject" (1922, p. 148). At the end of the addendum to his essay, he writes further,

> It remains to be established just what is the significance of the phrase "compatible with the constraints" which, in the presentation here given, modifies the maximum principle enunciated. The present communication is intended rather a preliminary than as an attempt to say the last word on the subject. (1922, p. 150)

The length of the whole essay, including the footnotes, is just over four pages. The fact that Lotka did not take the time in this brief paper to "fully" explain all the different constraints to which systems are subject clearly does not demonstrate that the PMEF he described is not modified by these constraints.

This same point applies to the MPP described by Odum and Pinkerton. To clarify Odum's perception of the MPP, I should point out that just about every time he refers to the principle, he cites Lotka as the source of it. He never, at any point, describes the differences between his view of it and Lotka's. He always characterizes his work as further developing Lotka's MPP. This by itself gives us reason to believe that Odum recognized that the MPP applied to natural systems "so far as compatible with the constraints" on them, but we have plenty of other evidence of this recognition. Odum and Pinkerton describe the optimum efficiency needed to maximize power increasing as the exergy decreases (1955, p. 341, 343). If, for example, a natural system has access to only an extremely limited amount of renewable energy, such as plants in Oregon in a partially shaded area in the winter, Odum and Pinkerton describe a higher level of efficiency being required in order to maximize useful power output. In this situation, power and efficiency will vary in a manner that is positively correlated.[5]

---

[5] The MPP can also be used to explain how the power and efficiency of a system can covary in a negative fashion that favors efficiency (as efficiency increases power decreases). If a natural system only has access to a diminishing quantity of nonrenewable energy, the power of the system will eventually begin to decrease (Odum, 1977a, p. 124, 126, Fig. 10b). In order to maximize the useful power output to the extent possible given these constraints, a system can increase its effi-

Odum does not describe the MPP as a "maximum power at intermediate effi-ciency principle"; he argues that according to the MPP, "systems are selected to operate at an *optimum* efficiency that generates maximum power" (my emphasis) and he describes how the optimum efficiency is influenced by the constraints acting on natural systems (Hall, 1995, p. XIII).

Glazier goes on to conclude that natural selection may give a selective advantage to power or efficiency depending on the resource quantity or quality and other con-straints on systems (2024, p. 36): when the resources are abundant, increased power will be favored; when resources are limited, increased efficiency will be favored. Once again, this is consistent with Odum and Pinkerton's description of the MPP. They describe how the intensity of light affects the efficiency of photosynthe-sis needed to maximize useful power output (1955, p. 341; Odum, 2007, p. 214). Odum suggests that the fact that the efficiency of photosynthesis is negatively cor-related with light intensity is because of the MEP (2007, p. 214). Furthermore, neither Odum nor Lotka suggest that the quantity of available energy is the only constraint on systems that affects the MPP or the PMEF. Odum and Pinkerton also suggested that the fact that the available light was flashing in the experiments done by Burk (1953) could in part explain the higher efficiencies of photosynthesis uncovered in his work: they implicitly suggest that the quality of available energy can influence the way power is maximized. This should be no surprise given all the work Odum goes on to do on the importance of energy quality.

Odum also describes how the prey-predator oscillations illustrated in the Lotka-Volterra model illustrate how the unique relation between prey and predator can cause the processes that maximize power to have a pulsing rhythm (2007, p. 155). This complex relation involves the stability of populations and the lethal threats posed by predators, two of the constraints Glazier discusses in his investigation (2024, p. 2; p. 33). For example, Glazier discusses what he calls the *deleterious effects of speed*: he argues that increasing the speed of various activities may increase the probability of injury, bad decisions, and exposure to predators, among other things. Glazier suggests that these effects may explain why productive effi-ciency often varies in a manner that is positively correlated with productive power, instead of the negative correlations "predicted by 'maximum power theory'" (2024, p. 33, 35). The evidence suggests that Odum would view the *deleterious effects of speed* described by Glacier as another constraint acting on natural systems that influences the way the MPP applies to them, like the quantity and quality of exergy.

In 1983, Odum acknowledged that there are instances in which many different criteria can be optimized in order to support the survival and growth of a natural

---

ciency. In this case, power and efficiency will covary in a negative fashion that favors efficiency—efficiency increases as power decreases—and it will do so in a manner that is consistent with the MPP because the system will maximize the useful power output that is "available" (Odum, 1982, p. 35). This is essentially the situation that Odum describes human civilizations entering after the peak of our ability to use fossil fuels (2001). Consequently, the MPP can explain how, under cer-tain conditions, power and efficiency can be negatively correlated in a way that favors power or efficiency, or they can be positively correlated (they go up or down together).

system: maximum efficiency, maximum entropy generation, minimum entropy generation, maximum diversity, maximum biomass, maximum reproduction, maximum profit, etc. He argues that they are all "special cases" of the MPP (1983a, p. 101, 1983b, p. 79). Here, Odum appears to recognize that when all of the constraints acting on systems that influence the way power is maximized are taken into account, the MPP can take the shape of many different criteria. The textual evidence is clear: Odum would view the many different constraints on systems Glazier discusses throughout his essay as a further development of Lotka's general idea that the MPP applies to natural systems in a way that is "compatible with the constraints" on these systems (see, in particular, the various intrinsic and extrinsic limiting factors Glazier discusses in Sect. 5.2). Glazier's analysis may provide an accurate critique of "maximum power theory," but it appears to provide an comprehensive illustration of the overlooked implications of the MPP: the impressive collection of data he gathered illustrates the many different ways the constraints on systems shape the processes that maximize power in different environments, giving rise to the spectacular complexity of nature.

In retrospect, we may question whether Odum and Pinkerton made a sufficient effort to make their position clear, whether they allowed the idealized example provided by Atwood's machine to dominate their description of the MPP to an extent that undermined the readers ability to fully understand the many different ways that useful power output is maximized within the context of the many different constraints on natural systems. One could however argue that this concern is unfair because their paper was so revolutionary: how were they supposed to know how it would eventually be misinterpreted? We may also recognize an evolution in Odum's own understanding of the MPP. In the "energy-rich culture" in the United States in the 1950s, Odum and Pinkerton wrote that "in natural systems, there is a general tendency to sacrifice efficiency for more power output" (1955, p. 331); near the end of his life, when he could see that after the peak of our ability to use fossil fuels the available energy would sharply decrease, Odum wrote a book about how societies will eventually need to become more efficient (2001). Here, we can see that throughout his career, Odum continued to explore the implications of the MPP and his own view of it evolved in a way that is consistent with the principle.

4.12.3   The MPP Vs. The MEPP

As mentioned in Sects. 4.9 and 1.1, scientists began using the maximum entropy production principle (MEPP) to explain natural phenomena instead of the MPP in the 1970s (Paltridge, 1975). This trend has continued, and now many scientists appear to prefer the MEPP (Sciubba, 2011, p. 1351). These scientists generally do not provide arguments to justify their move from the MPP to the MEPP, but this migration could, in itself, be considered an implicit argument against the MPP.

Charles Hall has identified an important problem with the MEPP which is crucial to any evaluation of the relative merits of the MEPP and the MPP (Hall & McWhirter,

2023, p. 7). He points out that Odum and Pinkerton argue that those natural systems and organisms that are able to direct more useful power output through the processes associated with survival and reproduction will have a selective advantage in evolutionary processes because they are likely to generate more surviving offspring than are those who do not (1955, p. 332). As a result, a higher rate of entropy production will also have a selective advantage, but, Hall argues, this advantage is only incidental, because of the fundamental relation between useful power and entropy production. Useful power output through the processes associated with survival and reproduction is, fundamentally, what has a selective advantage in evolutionary processes. Consequently, notwithstanding the present popularity of the MEPP, the development of natural systems through evolutionary processes appears driven fundamentally by the MPP. So far, no scientist who uses the MEPP has addressed this issue.

4.13   The MPP and Morality, Religion and Politics

As a systems ecologist, Odum could not ignore human social systems and their relation to ecological systems. Furthermore, he understood that the laws of thermodynamics applied to human social systems as well as ecological systems: for example, these systems too were products of evolutionary processes and the MPP. This section describes Odum's view of the power manifested by human social systems and the moral and religious values that influence their development. I discuss this aspect of Odum's work at the end of this chapter because I will be discussing Nietzsche's concept of the will to power and its relation to his critique of morality in the next chapter, and I will be highlighting the parallels and differences in their views.

4.13.1   Human Power

Traditionally, the concept of power as used in the natural sciences is not taken to apply directly to human social systems. Odum does not agree with this view. He begins Chap. 3 of *Environment, Power, and Society for the Twenty-First Century* discussing the concept of power used in the natural sciences and its relation to the forms of power manifested by human beings. He describes how in science and engineering, power is defined precisely in terms of measurable units, like "the calorie, the joule, the British thermal unit (Btu), and the erg" (2007, pp. 32–33). The forms of power manifested by human beings, on the other hand, are often considered to be different from the forms of power analyzed in the natural sciences. However, Odum goes on to point out that human forms of power can be measured in the same way.

> In human affairs the word "power" usually refers to the effectiveness of action or the capability of action. Great military power implies large military bodies involving many people and machines and exerting a directing force over large areas. Great political power suggests

> command of large numbers of votes and a wide influence on government systems and on the actions of many people. Great economic power implies control of large amounts of money and of influences that can be bought with volume spending. Almost everyone understands in a qualitative way what power means in human affairs, but few equate general concepts of power with scientific measures... Although nearly everyone is familiar with power ratings of household appliances and automobiles, our educational system has rarely emphasized that the affairs of people also have quantitative power ratings and that the important issues of human existence and survival are as fully regulated by the energy used and the laws of energetics as are the machines. It is possible to put calories-per-day values on human institutions, on the flows of energy in cities, on the power needs and delivery of activities of nations, or on the relative influences exerted by humans and their environmental systems. (2007, p. 32–33)

Odum believed we are able to compare the ability of all natural systems to do work, including human social systems.

He noted that, "our educational system" does not tend to focus on how human affairs are as "fully regulated by the laws of energetics as are the machines." These forms of human power are generally considered by scientists to be different from the forms of power that are described by the concept of power used in the natural sciences. As a systems ecologist, Odum views human social systems as parts of nature like ecological systems and he believes the scientific concept of power can be used to understand both these kinds of systems.

4.13.2 Moral and Religious Values

Odum sees human beings as products of an evolutionary process that is governed by thermodynamic laws, and the moral and religious values used by human beings are products of this same process. He argued that if we examine the flows of energy that produce the natural systems on the planet and the laws that guide them, we can begin to see how human values are related to these flows of energy. One of the kinds of information that has the highest energy quality are the moral and religious values that command the attention of large numbers of people. These values can help reproduce and guide the development of social systems, like the seeds of trees can reproduce and guide the development of more trees.

Odum believed that human beings have the freedom to make individual choices (1977a, p. 111). This, in fact, is an important part of the process of cultural evolution for human social systems; it enables adaptations to emerge faster than through the process of biological evolution alone (1977a, p. 114). Individual choices can play a role in cultural evolution that is similar to the role of genetic mutations in biological evolution: they both provide a range of options that can be selected for or against through evolutionary processes.

Individuals often believe the moral and religious values they follow are justified by their belief in some view of God, religion, philosophy, or science. Odum argues that in the process of cultural evolution, which includes our free choices, the moral and religious values needed to maximize power will have a selective advantage

(1977a, p. 114).[5] In a section entitled "Roots of ethics in energy laws and energy principles," he writes: "Ecological and other systems that survive and prosper have characteristic patterns adapted to the realities of energy laws and ecological principles" (Odum, 1977b, 175). He argues that human social systems are "subject to the same energy constraints as" other systems and, consequently, the moral and religious values associated with human social systems "must meet the same requirements." He concludes that, "according to this analysis culture is formed by the trial, selection, and survival process that produces a structure of information that tends to retain successful patterns, at least until conditions change." The particular moral and religious values we follow now are products of this evolutionary process.

Scientists developed our knowledge of cultural evolution further in substantial ways over the course of Odum's long career. For a thorough and authoritative discussion of our present knowledge of cultural evolution, see Joseph Henrich's book *The Secret of our Success* (2016). Just to provide a few examples of the processes associated with intergroup competition in cultural evolution, consider the following excerpts from this book: *"War and raiding"*—cultures that are better able to work together and/or develop the military, technological, and economic ability to defend themselves will be better able to prevail in military conflicts and pass their moral values on to the next generation; *"Differential group survival without conflict"*—societies that have the ability to survive, grow, and flourish more than others will be able to sustain a larger population and will therefore have more families with parents that share their moral values with their children; *"Differential migration"*—these societies that are able to grow and flourish more than others will be able to attract and support more immigrants from other societies who will learn to follow the moral values prevalent in the society; *"Prestige-biased group transmission"*—members in other societies will be more likely to imitate the behavior of people in societies that have been able to grow and flourish because they are often perceived to be more "prestigious" (Henrich, 2016, p. 167–168). Other principles also play a role, like skill-bias (p. 39), success-bias (p. 39), and conformist transmission (p. 47). These processes and principles illustrate how those societies that maximize power also maximize their power to share their moral and religious values with individuals in future generations.

Odum notes that many people believe that human intentions, such as the intention to be faithful to a particular religion, are the cause of the behavior of human social systems (1977a, p. 111). Scientists researching cultural evolution give us reason to think a larger process is at work that includes human intentions. For example, scientists found that women living in primitive tribes in Fiji adhere to a series of food taboos during pregnancy and breast-feeding that selectively remove the most toxic marine species from their diet (Henrich, 2016, p. 99–101). During this period, women have a reduced resistance to toxins; these taboos therefore substantially reduce fish poisoning. When the women in these tribes were asked why they follow these taboos, they often said that they did not know and they appeared to think it was an odd question. Others simply said that it was "custom." Scientists explain these taboos as being the product of cultural evolution: those cultures that are the healthiest and are able to flourish are better able to pass their customs on to others. Some scientists argue that the process of cultural evolution has a kind of "wisdom" that can enable a social system to flourish which is not completely understood by the humans in the system. These scientists do not believe the intentions of their subjects are the only causal factor that determines their behavior; they view the intentions themselves as products and signs of a powerful evolutionary process. Lotka suggested that it now appears probable that human beings have "been unconsciously fulfilling a law of nature" through this evolutionary process (1922, p. 149); Odum argues that the law of nature is the MPP, and they both argue that our moral and religious values are products of this process.

Odum describes the development of social systems being fundamentally structured by the energy available to them and he argues that this, in turn, influences the moral and religious values used in these systems. He writes that his general hypothesis is that "culture evolves to fit the energy pattern" (1977a, p. 128). If a social system has access to a steady source of renewable energy, the system will initially be able to grow, but the growth will eventually stabilize into a steady state (1977a, p. 124, Fig. 10). If a social system has access to a non-renewable source of high-quality energy, the system will initially be able to grow, but that growth will eventually taper off and the system will begin to decline as the amount of available energy declines. If a social system has access to both these sources of energy, it will initially be able to grow, but this growth will eventually begin to taper off and the system will begin to decline; eventually, this decline will taper off and the system will reach a steady state fueled by the renewable source energy. The fact that the human population of the planet was essentially stable for millennia before the industrial revolution is consistent with this view.

When social systems are able to grow, moral values will evolve to foster and sustain this growth. There will be a faith in progress and the moral values and attitudes associated with it. The conservative values associated with social systems in a steady state will no longer apply. Odum argues that excessive sexual promiscuity would present a threat to societies that have access to a limited amount of energy and need to use it as efficiently as possible; in this context, the behavior would generally be considered immoral. But if energy is abundant and the society is growing, such behavior does not present an important threat and it would not be considered immoral. Odum suggests that the fact that president Clinton was not removed from office in the United States because of his sexual indiscretion is evidence of this more forgiving attitude toward sexual promiscuity during a period when energy was abundant (2007, p. 327). The United States in the twentieth century provides a good example of a society system during a period of accelerated growth.

When social systems are in a steady state, moral values will evolve in a direction that does not place a high value on progress, change, or growth. A high value will be placed on efficiency, unity, and understanding our role in the biosphere. Individuals with moral values that are consistent with this focus will be able to generate more power in these societies. Odum uses Japan's rural culture prior to the development of its industrial economy as an example of a society in a steady state (1977a, p. 127–131). This culture used much less energy, was relatively simple, developed over a long period, and developed in a way that was closely connected with its environment. A fundamental element of this culture was that it used the water from the mountains to support the rice paddies. Soybeans were an important part of the diet; they used nitrogen from the air to fertilize the soil. The Shinto religion that was practiced by many has symbols, faith, and fears connected with the forest. It was believed that the purity of the waters kept the human beings in harmony with the forest and nature.

When social systems are declining, it can cause many people to become dissatisfied which can lead them to seek security in the religious and moral values of the past. Others will be more willing to consider new values that are better able to address the new conditions and challenges. Political differences can emerge between these two approaches, like the differences between conservative and liberal approaches to political issues. Odum argues that moral and religious values will evolve in a way that no longer focuses on consuming luxuries and aggressively competing for more money; they will eventually come to focus on education, location efficient lifestyles, relationships that absorb sexual energy without leading to reproduction, birth control, efficient infrastructure, personal responsibility for health (healthy diet, regular exercise, practicing safe sex), and respect for diversity. Individuals that have values that focus on these things will be able to generate more power in these societies.

Like many scientists today, Odum believed that contemporary human societies will begin to decline soon. The growth of these societies since the nineteenth century has been propelled by the use of fossil fuels, a nonrenewable form of energy, and the peak of our ability to access and use these fuels is approaching. On the other side of this peak, according to Odum's analysis, industrial societies will begin to decline. This led Odum to write a book entitled *A Prosperous Way Down* (2001), where he describes in detail the transition he sees coming in industrial societies, including the evolution of their moral and religious values.

In 2007, an updated version of Odum's book *Environment, Power, and Society* was published in which he provided an "Energy Systems Ethics for All Scales" that applies to societies at all stages of their development: growth, steady state, or decline. It includes the following: "seek satisfaction in useful contribution; help maximize real wealth (empower); reinforce environmental sources; treasure genetic and cultural diversity; adapt to natural hierarchy; minimize luxury; minimize waste; adapt to system rhythm; share information; optimize efficiency; circulate materials; circulate money; fit the earth; reproduce only as needed; have faith in self-organization" (2007, p. 329).

4.14 Conclusion

Howard T. Odum was one of the most prolific and creative scientists in the twentieth century. His earlier studies focused on understanding nature in terms of systems driven by energy. He went on to further develop Lotka's PMEF in a number of fundamental ways and eventually renamed it the Maximum Power Principle. This principle served as the centerpiece of Odum's systems view of ecology: it helps synthesize the way different parts of systems interact, and connects thermodynamics with biology, ecology, and economics. Odum and Pinkerton described a unique relation between the efficiency and speed of energy transformations that generates maximum useful power output and varies in a manner that is based on the constraints acting on systems. They described the MPP maximizing power *output*; Lotka, on the other hand, described the PMEF as maximizing the energy *flux* through natural systems. They also described the MPP maximizing *useful* power output; Lotka did not make such a distinction. Odum saw feedback loops playing an integral role in the Darwinian processes that maximize useful power output. He described useful power output being maximized in natural systems over time through a pulsing rhythm. Later, he began to focus on the importance of the quality of energy and this eventually led him to use the concept of emergy and rename the MPP the *maximum empower principle*. In the 1970s and 80s, many contemporary scientists began to use a principle that is closely related to the MPP called the maximum entropy production principle.

This chapter reviewed the empirical evidence that supports the MPP, including evidence that has emerged over the past few decades. It also discussed the theoretical support for the MPP. This chapter critically reviewed some arguments against the MPP and the maximum empower principle that have been offered by contemporary scientists. It provides an argument for a non-reductive interpretation of the MPP that recognizes that the processes that maximize useful power output are fundamentally influenced by the different constraints on natural systems, and this can alter, among other things, the relative value of efficiency. This chapter also provides a brief but important argument regarding the relative value of the MPP and the maximum entropy production principle.

This chapter ends with a discussion of how Odum described the MPP guiding the evolution of moral and religious values. In the previous chapter, we discussed how Lotka critically analyzed the behavior of human beings from the perspective of his PMEF. In the next chapter, I will discuss Nietzsche's concept of the will to power, how he used it to critically evaluate moral and religious values, and some of the philosophical implications of his approach.

References

H. T. Odum's Works

Odum, H. T. (1960). Ecological potential and analog circuits for the ecosystem. *American Scientist*, *48*, 1–8.
Odum, H. T. (1971). *Environment, power, and society*. Wiley.
Odum, H. T. (1973). Energy, ecology and economics. *Royal Swedish Academy of Science. AMBIO*, *2*(6), 220–227.
Odum, H. T. (1977a). The ecosystem, energy, and human values. *Zygon, 12*(2), 109–133.

Odum, H. T. 1977b. Energy, value, and money. In Ecosystem modeling in theory & practice, eds. Charles A. S. Hall and John W. Day, Jr., 173–196. : Wiley.

Odum, H. T. (1982). Pulsing, power and hierarchy. In W. J. Mitsch, R. K. Ragade, R. W. Bosserman, & J. A. Dilon (Eds.), *Energetics and systems* (pp. 33–60). Ann Arbor Science Publishers.

Odum, H. T. (1983a). *Systems ecology: An introduction.* Wiley.

Odum, H. T. (1983b). Maximum power and efficiency: A rebuttal. *Ecological Modeling, 20*(1), 71–82.

Odum, H. T. (1995). Self-organization and maximum empower. In C. A. S. Hall (Ed.), *Maximum power: The ideas and applications of H. T. Odum.* University Press of Colorado.

Odum, H. T. (2007). *Environment, power, and society for the 21th century.* Columbia University Press.

Odum, H. T., & Hoskin, C. M. (1958). Comparative studies of the metabolism of Texas bays. *Publications of the Institute of Marine Science, University of Texas, 5,* 16–46.

Odum, H. T., & Odum, E. C. (1976). *Energy basis for man and nature.* McGraw-Hill.

Odum, H. T., & Odum, E. C. (2001). *A prosperous way down.* University Press of Colorado. Kindle version.

Odum, H. T., & Pigeon, R. F. (Eds.). (1970). *A tropical RainForest; a study of irradiation and ecology at El Verde, Puerto Rico.* United States Atomic Energy Commission, National Technical information service.

Odum, H. T., & Pinkerton, R. (1955). Time's speed regulator: The optimum efficiency for maximum power output in physical and biological systems. *American Scientist, 43*(2), 331–343.

Odum, H. T., & Richardson, J. R. (1981). Power and a pulsing production model. In W. J. Mitsch, R. W. Bosserman, & J. M. Klopatek (Eds.), *Energy and ecological modelling: Developments in environmental modeling 1* (pp. 641–648). Elsevier Scientific Publishing. (CFW-81-23).

Odum, H. T., Slier, W. L., Beyers, R. J., & Armstrong, N. (1963). *Experiments with engineering of marine ecosystems.* Public of the Institute of Marine Science, University of Texas, 9:374–403.

Odum, H. T., Odum, W. E., & Odum, E. P. (1995). Nature's pulsing paradigm. *Estuaries and Coasts, 18,* 547–555.

Other Sources

Ameniya, T., Itoh, K., & Serizawa, H. (2011). Principle of maximum entropy production –applicability to social analysis. *Sociological Theory and Methods, 26*(2), 405–420.

Atkinson, Q., Meade, A., Venditti, C., Greenhill, S., & Pagel, M. (2008). Languages evolve in punctuational bursts. *Science, 319*(5863).

Bonanno, A., & Reuter, M. (2011). Entropy production during asymptotically safe inflation. *Entropy, 13*(1), 274–292.

Brown, M. T. (2023). The maximum power principle. In E. P. Rosa & J. Ramos-Martin (Eds.), *Elgar encyclopedia of ecological economics* (pp. 363–367). Edward Elgar Publishing.

Brown, M. T., & Ulgiati, S. (2004). Energy quality, emergy, and transformity: H.T. Odum's contributions to quantifying and understanding systems. *Ecological Modeling, 178,* 201–213.

Burk, D. (1953). Photosynthesis: A thermodynamic perfection of nature. *Federation Proceedings, 12*(2), 611–625.

Cai, T. T., Olsen, T. W., & Campbell, D. E. (2004). Maximum (em)power: A foundational principle linking man and nature. *Ecological Modeling, 178,* 115–119.

Cai, T. T., Montague, C. L., & Davis, J. S. (2006). The maximum power principle: An empirical investigation. *Ecological Modeling, 190*(3–4), 317–335.

Cevolatti, D., & Maud, S. (2004). Realizing the enlightenment: H. T. Odum's energy systems language *qua* G. W.V Leibniz's *Characteristica Universalis. Ecological Modeling, 178,* 279–293.

Cook, E. (1976). *Man, energy, society.* W. H. Freeman.

Cordaro, E. G., & Venegas-Aravena, P. (2024). The multiscale principle in nature (*principium luxuriae*): Linking multiscale thermodynamics to living and non-living complex systems. *Fractal and Fractional, 8/35*. https://doi.org/10.3390/fractalfract8010035

Court, V., & Fizaine, F. (2017). *Long-term estimates of the energy-return-on-investment (EROI) of coal, oil, and gas global productions*. University of Sussex. Journal contribution. https://hdl.handle.net/10779/uos.23461727.v1

Dewar, R. (2003). Information theory explanation of the fluctuation theorem, maximum entropy production and self organized criticality in non-equilibrium stationary states. *Journal of Physics A: Math and General, 36*, 631–641. https://doi.org/10.1088/0305-4470/36/3/303

Dewar, R. (2005). Maximum entropy production and the fluctuation theorem. *Journal of Physics A: Math and General, 38*, L371–L381. https://doi.org/10.1088/0305-4470/38/21/L01

Eldredge, N., & Gould, S. J. (1972). Punctuated equilibria: An alternative to phyletic gradualism. In T. J. M. Schopf (Ed.), *Models in paleobiology* (pp. 82–115). Freeman Cooper.

Eldredge, N., & Gould, S. J. (1993). Punctuated equilibrium comes of age. *Nature, 366*, 223–227.

England, J. (2013). Statistical physics of self replication. *The Journal of Chemical Physics, 139/12*, 121923/1–121923/8.

Farajollahi, H., Salehi, A., & Tayebi, F. (2011). The generalized second law in chameleon cosmology. *Canadian Journal of Physics, 89*(9), 915–919.

Fleck, G., & Morel, R. (2006). A fourth law of thermodynamics. *Chemistry, 15*(4), 305–310.

Fry, I. (1995). Evolution in thermodynamic perspective: A historical and philosophical angle. *Zygon, 30*(2), 227–248.

Fry, I. (2000). *Emergence of life on earth: A historical and scientific overview*. Rutgers University Press.

Glacier, D. S. (2024). Power and efficiency in living systems. *Science, 6*(28).

Hall, C. A. S. (1995). Introduction: What is maximum power? In C. A. S. Hall (Ed.), *Maximum power: The ideas and applications of H. T. Odum*. University Press of Colorado.

Hall, C. A. S. (2004). The continuing importance of maximum power. *Ecological Modeling, 178*, 107–113.

Hall, C. A. S. (2017). *Energy return of investment: A unifying principle for biology, economics, and sustainability*. Springer.

Hall, C. A. S., & McWhirter, T. (2023). Maximum power in evolution, ecology and economics. *Philosophical Transactions of the Royal Society A, 381*, 20220290. https://doi.org/10.1098/rsta.2022.0290

Hall, C. A. S., Cleveland, C. J., & Kaufmann, R. (1986). Energy and resource quality. *The Ecology of the Economic Process*. Wiley.

Hall, C. A. S., Harris, N. L., & Lugo, A. E. (2013). A test of the maximum power hypothesis along an elevational gradient in the Luquillo Mountains of Puerto Rico. *Ecological Bulletins, 54*, 233–243.

Henrich, J. (2016). *The secret of our success*. Princeton University Press.

Herendeen, R. A. (2004). Energy analysis and EMERGY analysis—A comparison. *Ecological Modelling, 178*, 227–237.

Hunt, G. (2008). Gradual or pulsed evolution: When should punctuational explanations be preferred? *Paleobiology, 34*(3), 360–377.

Jorgensen, S., & Svirezhev, Y. (2004). *Towards a thermodynamic theory for ecological systems*. Pergamon.

Kleidon, A., & Lorenz, R. D. (2005). *NET and the production of entropy: Life, earth and beyond*. Springer.

Kleidon, A., Malhi, Y., & Cox, P. (2010). Maximum entropy production in environmental and ecological systems. *Philosophical Transactions of the Royal Society of Biological Sciences B, 365*, 1297–1302.

Kuhn, T. (1962). *The structure of scientific revolutions*. University of Chicago Press.

Langbein, W. A., & Leopold, B. (1962). *The concept of entropy in landscape evolution*. U. S. Geological Survey Professional Paper 550A.

Lenton, T. M., Pichler, P., & Weiz, H. (2016). Revolutions in energy input and material cycling in Earth history and human history. *Earth System Dynamics, 7*(2), 353–370.

Lotka, A. (1921). Note on moving equilibria. *Proceedings of the National Academy of Sciences of the United States of America, 7*(6), 168–172.

Lotka, A. (1922). Contribution to the energetics of evolution. *Proceedings of the National Academy of Science, 8*(6), 147–151.

Lotka, A. (1945). The law of evolution as a maximal principle. *Human Biology, 17*(3), 167–194.

Martyushev, L. M., & Seleznev, V. D. (2006). Maximum entropy production principle in physics, chemistry and biology. *Physics Reports, 406*(1), 1–45.

Michaelian, K. (2011). Entropy production and the origin of life. *Journal of Modern Physics, 2*(6), 595–601.

Nagel, T. (2012). *Mind and cosmos: Why the materialist neo-Darwinian conception of nature is almost certainly false.* Oxford University Press.

O'Neill, R. (1996). *Maximum power: The ideas and applications of H. T. Odum* by Charles A. S. Hall. *Ecology, 77/7*, 2263.

Paltridge, G. W. (1975). Global dynamics and climate? A system of minimum entropy exchange. *Quarterly Journal of the Meteorological Society, 104*, 927–945. https://doi.org/10.1002/qj.49710444206

Prigogine, I., & Stengers, I. (1984). *Order out of chaos: Man's new dialogue with nature.* Bantam Books.

Sciubba, E. (2009). Why emergy- and exergy analyses are non-commensurable methods of system analysis. *International Journal of Exergy, 6/4*, 523.

Sciubba, E. (2011). What did Lotka really say? A critical reassessment of the "maximum power principle". *Ecological Modeling, 222*, 1348–1353.

Silvert, W. (1982). The theory of power and efficiency in ecology. *Ecological Modeling, 15*, 159–164.

Smith, C. C. (1976). When and how much to reproduce: The trade-off between power and efficiency. *American Zoologist, 16*, 763–774.

Steffen, W., Broadgate, W., Deutsch, L., Gaffney, O., & Ludwig, C. (2015). The trajectory of the anthropocene: The great acceleration. *The Anthropocene Review, 2*(1), 81–98.

Sugita, M. (1951). Maximum principle in transient phenomena and its application to biophysics. *Bull. Kobayasi Institute, 1*, 88–101.

Swenson, R. (1989). Emergent attractors and the law of maximum entropy production: Foundations to a theory of general evolution. *Systems Research, 6*, 187–197.

Thimsen, E. (2024). Planetary energy flow and entropy production rate by Earth from 2002 to 2023. *Entropy, 26*(5), 350.

Ulanowicz, R. E., & Hannon, B. (1987). Life and the production of entropy. *Proclamations of the Royal Society of London, 232*, 181–192. https://doi.org/10.1098/rspb.1987.0067

Chapter 5
Nietzsche's Will to Power

Abstract If we look for concepts used by philosophers that relate to the maximum power principle, there is one that clearly stands out: Nietzsche's concept of the will to power. It holds that all natural systems, both biotic and abiotic, develop in ways that increase their power. After his death in 1900, Nietzsche's philosophy went on to have a substantial impact on twentieth and twenty-first century philosophy and cultures around the world. However, most of the philosophers that have reviewed his work have concluded that this interpretation of the will to power is empirically implausible. This chapter reviews Nietzsche's description of the will to power and the argument he provides for it; it also illustrates how this concept plays a central role in Nietzsche's critique of morality and then provides a unique interpretation of this critique. Over the past few decades, a number of philosophers have offered new and influential interpretations of Nietzsche's critique of morality: some have argued that he uses the will to power as a fundamental moral value that has a privileged normative status; others have argued that the status of the will to power is not privileged in any way. It is just the perspective he chooses to use. This chapter argues that Nietzsche does not claim the moral values he analyzes are either immoral or moral, he *describes* these values as fostering or undermining the growth and flourishing of life as outlined by the will to power, understood as an empirical principle: his critique of morality comes from the perspective of science. The will to power is privileged because it is better able to explain natural phenomena than alternative theories. The interpretation of Nietzsche that emerges brings into relief the extent to which Odum's critique of moral and religious values provides a model for Nietzsche's critique of morality.

5.1 Introduction

If we look for the philosophical roots of the maximum power principle (MPP), even a cursory review of the history of philosophy will bring us to a particular concept: Nietzsche's will to power. Thirty six years before Lotka argued that natural selection was guided by the principle of maximum energy flux, and sixty nine years

© The Author(s), under exclusive license to Springer Nature Switzerland AG 2025
T. McWhirter, *Maximum Power and its Philosophical Roots*, SpringerBriefs in Energy, https://doi.org/10.1007/978-3-031-80622-3_5

before Odum and Pinkerton argued that natural selection was guided by the MPP, Nietzsche argued that the evolution of all natural systems was guided fundamentally by a drive to manifest more power (*BGE* 13, 36).[1] This drive is its own end; it has no other purpose or goal: it is a will to power [*Wille zur Macht*]. He saw it applying to physics, biology, psychology, and sociology (*BGE* 6, 12, 13, 19, 36).[2] He writes: "The world viewed from inside, the world defined and determined according to its "intelligible character"—it would be "will to power" and nothing else.—" (*BGE* 36).

Nietzsche's philosophy and his concept of the will to power, like the work of many philosophers and scientists in his time, can be viewed as the products of an effort to deal with two major revolutions in human understanding that took place in the nineteenth century: Darwin's evolutionary theory and the laws of thermodynamics. Initially, many scientists and philosophers did not see evolutionary theory and the second law of thermodynamics as being compatible: how can the organization of life develop over time if entropy increases? Nietzsche rarely mentions Darwin, and when he does, it is usually to criticize some aspect of his work (*TI* Skirmishes 14); nonetheless, Nietzsche accepts basic aspects of Darwin's view and many aspects of Nietzsche's philosophy are developed in response to his criticism of Darwin. Darwin's influence on Nietzsche is outlined in considerable detail in John Richardson's book *Nietzsche's New Darwinism* (2004). Richardson argues that, according to Nietzsche, organisms do not increase their power because they intend to do so; the power of organisms tends to increase because organisms that are more powerful have a selective advantage in evolutionary processes (2000, p. 112).

Nietzsche was also aware of the second law of thermodynamics; he discusses William Thomson's (1824–1907) interpretation of the cosmological implications of the second law in his notes (*WP* 1066).[3] In retrospect, we can now see Nietzsche's concept of the will to power as a prescient effort to explain how evolution and the second law are compatible, but, in his time, the concept was ignored by philosophers and scientists. After his death in 1900, many aspects of his philosophy became more popular for many different reasons: he had an influence on the development of existentialism, postmodernism, and psychoanalysis; his critique of morality has had lasting impact on discussions of moral theory; his critique of Christianity and religions liberated many from a blind obedience to religious faith; and his philosophy has had an extraordinary impact on cultures around the world, which is reflected in

---

[1] Translations of Nietzsche's quotes refer to the editions listed in the references. The standard abbreviations listed in the references section will be used, with the Arabic numerals referring to aphorisms and Roman numerals referring to sections. References to Nietzsche's unpublished works included in the *KSA* will include the term Nachlass, the year of the note, the fragment number, *KSA*, the volume number, a decimal point, and then the page numbers (e.g., Nachlass 1888, 14[121], *KSA* 13.300–301).

[2] See also, Nachlass 1888, 14[121], *KSA* 13.300–301.

[3] Nachlass 1888, 14[188], *KSA* 13:374–76.

music (Richard Strauss's *Thus Spake Zarathustra*), numerous books and movies (Stanley Kubrick's *2001: A Space Odyssey*), and common sayings ("what does not kill me, makes me stronger"). There are now three philosophy journals that focus exclusively on his work and more and more books on his philosophy are published every year. However, his concept of the will to power has always been controversial, and the version of it that applies to biotic and abiotic systems has almost universally been viewed by philosophers as being implausible according to the contemporary sciences. This will be discussed in Sect. 5.2.3.

This chapter focuses on explaining the will to power and the argument that Nietzsche provides for it. It also explains how the will to power relates to Nietzsche's critique of morality and develops an argument for a unique interpretation of this critique. Given the similarities between the MPP and the will to power and Odum's and Nietzsche's use of them to critically analyze moral and religious values, this explication of the philosophical implications of Nietzsche's critique of morality will also help illustrate the philosophical implications of Odum's critique of moral and religious values.

Some philosophers argue that Nietzsche's critique of morality makes moral claims that can be justified because the will to power is a fundamental moral value. Other philosophers argue that the moral claims Nietzsche makes in his critique of morality cannot be justified because he does not provide a compelling argument that the will to power is itself a fundamental moral value. This investigation provides an argument that Nietzsche does not criticize moral values for being immoral; he describes how moral values affect the growth and flourishing of life as outlined by the will to power, understood as an empirical principle. The will to power provides a thermodynamic framework for an empirical theory of social development which supports Nietzsche's descriptions of the effects of moral values. He sees this approach as a "science of morality" (*BGE* 186). Let me go ahead and acknowledge at the outset that Nietzsche's critique of morality does not appear to be all that scientific, particularly from the perspective of contemporary scientists. His style of writing and thinking is unsystematic, disorganized, and unusual. It would be more accurate to describe his critique of morality as outlining the foundations of a "science of morality" that can be further developed in the future by someone like H. T. Odum.

This chapter, more than any other, therefore will explain the *philosophical* roots of the MPP. It has two major parts: the first part focuses on explaining the will to power and the argument Nietzsche provides for it; the second part explains my argument for a unique interpretation of Nietzsche's critique of morality and how it relates to other influential interpretations.

5.2   The Will to Power

5.2.1   Nietzsche's Description of the Will to Power

One of Nietzsche's first references to the will to power is in *Thus Spake Zarathustra,* where Zarathustra makes it clear that the will to power is a fundamental part of life "*as it is:*" "Where I found the living, there I found will to power; and even in the will of those who serve I found the will to be master.;" "Only where there is life is there also will: not will to life but—thus I teach you—will to power." (Z II "On Self-overcoming"). Nietzsche discusses the will to power in more detail in *Beyond Good and Evil,* where he describes it as an empirical principle, rather than an *a priori* metaphysical principle or a fundamental moral value. He argues, "*Physiologists* should think before putting down the instinct of self-preservation as the cardinal instinct of an organic being. A living thing seeks above all to *discharge* its strength—life itself is *will to power*;[…]" (*BGE* 13; the emphasis on *physiologists* is mine). The contemporary philosopher John Richardson suggests that we can better under-stand the will to power if we acknowledge that Nietzsche accepts many aspects of Darwin's theory of biological evolution, he just argues that this evolution is not guided by a struggle for "self-preservation;" it is guided by the will to power (2004, p. 11, 18; *TI* "Skirmishes" 14). In *The Gay Science,* he suggests "*natural scientist[s]*" should recognize the "will to power" is "the will to life" (*GS* 349; my emphasis).

While Nietzsche's first references to the will to power characterize it as a biologi-cal principle, he goes on to describe it as applying to abiotic and biotic systems in his published writings and his notes.[4] He describes it as applying to "a more primi-tive form of the world of affects in which everything still lies contained in a power-ful unity before it undergoes ramifications and developments in the organic process …—as a pre-form of life." (*BGE* 36). Here, he refers to it applying to pro-cesses that take place before they develop into an "organic process." In his notes, he writes: "what has been the relation of the total organic process to the rest of nature? That is where it's fundamental will stands revealed."[5] He sees the will to power as this fundamental will that organic processes share with the "rest of nature" (*WP* 691). He argues that, "The world" is the "'will to power' and nothing else" (*BGE* 36). He therefore views the will to power as applying to abiotic systems as well as biotic systems, and in this sense, the will to power is similar to the MPP and the maximum entropy production principle. Ontology is a branch of philosophy that studies the nature of being and existence and examines what all entities or natural systems have in common. Philosophers often refer to the view that the will to power refers to abiotic systems as well as biotic systems as the *ontological version of the will to power* (OWP).

---

[4] See *BGE* 22, 36; *GM* II 12; Nachlass 1884–1885, 35[15], *KSA* 11.513–514; Nachlass 1887–1888, 14[81–82], *KSA* 13.260–262; Nachlass 1884–1885, 38[12], *KSA* 11.610–611.

[5] Nachlass 1885–1886, 2[99], *KSA* 12.109.

In physics, power is defined as a measure of the rate at which energy is expended over time. Energy is defined as an ability to do work, like moving an object from one place to another. Power, therefore, is a measure of the rate at which things are transformed. Nietzsche describes the OWP as a drive in natural systems to manifest power that is transformative in at least three different senses: (1) It transforms the nature of systems themselves; (2) it transforms the environment of systems; (3) it transforms energy from a system's environment in a manner that enables the system to sustain and develop its organization and power.

First, the OWP refers to the tendency of systems to transform their own structure. Nietzsche describes this tendency giving rise to the development of "organs and functions with the disappearance of the intermediate members" (Nachlass 1885–1887, 7[9], *KSA* 12.297; *WP* 644). As a social process, it is associated with a "change of values" that is also a "change of creators" (*Z* I "On the Thousand and One Goals"). Nietzsche describes the OWP transforming social systems through the sublimation or spiritualization of the ends of the different drives operating on systems (Richardson, 1996, p. 321).[6] The sex drive, for example, can be sublimated into a form of love (*BGE* 189); the drive to manifest and enjoy cruelty can be sublimated into an enjoyment of bullfighting or the opera (*BGE* 229; *GM* II: 7). The will to power seeks to continually improve the ability to seek different ends. Nietzsche believed the ends themselves are not as important as the pursuit of them (1996, p. 341–343): he writes, "in the end one loves one's desire and not what is desired" (*BGE* 175). The will to power is not a drive that merely repeats itself, seeking the same thing more effectively; it eventually transforms the whole pattern of activity, raising it to a "higher" level. The will to power does not, for example, lead human beings to continually improve their ability to live as hunters and gatherers; it leads them to periodically transform their whole mode of production, raising it to a higher level. Obviously, in order for natural systems to come into being and increase their power, their structure must be transformed. Later, we will consider how this aspect of Nietzsche's view relates to Odum's pulsing paradigm.

Second, the OWP refers to the tendency of systems to transform other systems in their environment. Nietzsche describes it as being "essentially a will to violate" that which opposes it (Nachlass 1887–1889, 14[79], *KSA* 13.258), "to thrust back all that resists its extension" (*WP* 636; Nachlass 1887–1889, 14[186], *KSA* 13.373); it strives to dominate or "become master over" other bodies. In doing so, the will to power exerts an influence over these bodies, causing a change in their behavior.

Third, Nietzsche describes natural systems as being sustained through the transformation of energy from their environment. When considering the link between the organic and inorganic, Nietzsche describes the will to power as a will to transform (Nachlass 1885–1887, 10[138], *KSA* 12.535) or use up "energy," to "appropriate, dominate, increase, grow stronger" (*WP* 689; Nachlass 1887–1889, 14[81], *KSA* 13.261). The energy used by centers of force is described as coming from their environment. All centers of force grow through the "appropriation and assimilation"

---

[6] See *BGE* 58, 189, 198, 209, 229, 252, 271.

of external forces "until at length that which has been overwhelmed has entirely gone over into the power domain of the aggressor and has increased the same […]" (*WP* 656; Nachlass 1885–1887, 9[151], *KSA* 12.424): the world is thus fundamentally competitive, aggressive, and agonistic; it "lives on itself, its excrements are its food" (*WP* 1066; Nachlass 1887–1889, 14[188], *KSA* 13.374).

The metaphor of "nourishment" Nietzsche uses here and elsewhere in his discussion of inorganic and organic processes is particularly insightful (see also *BGE* 36): it highlights the fact that Nietzsche believed natural systems use energy from their environment to *sustain* their development. This is what happens when human beings acquire and eat food. Richardson notes that Nietzsche refers to hunger as a unique form of the OWP, which emerged from it through a division of labor; the OWP came to focus on eating because it is the way that organisms *assimilate* and *incorporate* external forces (*WP* 651; Nachlass 1887–88,11[121], *KSA* 13; Richardson, 1996, location 307). Nietzsche describes the will to power providing the solution to the problem of "nourishment" (*BGE* 36).

This third sense of transformation to which Nietzsche refers in the passages above illustrates that he had a prescient understanding of the dissipative nature of natural systems. The physicist Erwin Schrödinger eventually resolved the perceived contradiction mentioned in the introduction between evolution and the second law. He argued that living things extract and expend a large amount of energy from their environment to build their own smaller structure within a limited area. Thus, the entropy of the entire system increases to compensate for a living thing's own organization in a manner that is consistent with the second law (1945). The chemist Ilya Prigogine began to call natural systems *dissipative structures* because they survive by capturing energy from their environment and dissipating it as heat as they undertake the basic processes of life (1961).[7]

Nietzsche describes the OWP being transformative in these three senses. Implicit in this description, we also find the view that the will to power emerges through an interaction among different drives, systems, organisms, and sources of power. In this sense, we can see that the OWP relates to Nietzsche's *perspectivism*, which is a fundamental aspect of his epistemology, i.e., his theory of knowledge. Nietzsche viewed all forms of knowledge as being based on particular perspectives that are partial and provisional; no perspective is complete and unchanging (*BGE* 6). Richardson argues compellingly that Nietzsche viewed his perspectivism as being a fundamental part of the will to power (1996, p. 285–287). When Nietzsche describes different perspectives, he characterizes them as different wills to power. He viewed power as being "individuated": different wills manifest their power in different ways from different perspectives. Richardson believed Nietzsche considered this way of viewing things as being one of his most important original contributions and he used this view of the will to power to justify his perspectivism (1996, p. 175).

---

[7] William Plank, a French professor, discusses the relation between the will to power and dissipative structures in his book *The Quantum Nietzsche* (1998).

Consequently, if we interpret Nietzsche's perspectivism accurately, it is not undermined by the OWP; it is justified by it.

Nietzsche also uses the OWP to refer to the power manifested by human beings (*GM* II 12, 18; *GM* III 14, 15, 18, 27). This is not something that has been questioned by philosophers in the secondary literature, but scientists are generally not accustomed to using the scientific concept of power in this way. Odum's use of the scientific concept of power to understand the power manifested by human beings is an exception to this rule (see, e.g., 2007, p. 32–33), which illustrates a unique parallel between the MPP and OWP.

5.2.2   Nietzsche's Argument for the Will to Power

In *Beyond Good and Evil*, Nietzsche outlines his well-known argument for the OWP. This argument is clearly not as compelling as Lotka's theoretical argument for the principle of maximum energy flux, or Odum's and Pinkerton's argument for the relation between efficiency, speed, and the MPP. We now have much better arguments and evidence to support the OWP, the MPP, and principle of maximum energy flux. The first reason Nietzsche's argument is important to this chapter is that it provides pivotal textual evidence that supports the argument that I am making for my interpretation of Nietzsche's critique of morality. As we will see, some philosophers have tried to argue that Nietzsche really did not endorse the OWP; this argument provides textual evidence that he does endorse it. The second reason this argument is important is that it illustrates that Nietzsche provided an argument for the OWP decades before Lotka or Odum provided their arguments for the principle of maximum energy flux or the MPP. I am not aware of any other philosopher providing this kind of argument. Nietzsche's argument for the OWP therefore provides evidence that he is the most definitive philosophical root for the MPP.

Nietzsche begins to develop this argument in *BGE* 12, where he describes "materialistic atomism" as being based on the "the belief 'in substance,' in 'matter,' in the earth-residuum and particle-atom:," and he suggests it is "one of the best refuted theories there are." He attributes its refutation to Roger Joseph Boscovich (1711–1787), the eighteenth century natural philosopher who argued that matter is made up of points without dimension that exert fields of force.

Nietzsche goes on in *BGE* 12 to argue that we also need to "declare war" on the "atomistic need" in other areas of inquiry, including psychology, where it is reflected in the "soul atomism": "the belief which regards the soul as something indestructible, eternal, indivisible, as a monad, as an atomon: [...]." He argues this belief "ought to be expelled from science!" It should be replaced with "new versions and refinements of the soul-hypothesis," such as "soul as subjective multiplicity," and "soul as social structure of the drives and affects." Nietzsche argues that Boscovich's relational view of matter can provide a more accurate model for the activity of the soul.

In aphorism 36, Nietzsche further develops his argument for the OWP. There, he begins by assuming his revision of the soul-hypothesis discussed above (*BGE* 12 and 16 through 19).

> Suppose nothing else were "given" as real except our world of desires and passions, and we could not get down, or up, to any other "reality" besides the reality of our drives—for thinking is merely a relation of these drives to each other:[…]

Thus, Nietzsche does not describe thinking in terms of a "soul atomism;" he describes it in terms of a relation of different drives, desires, and passions: a "social structure of the drives and affects." He then goes on to write,

> Is it not permitted to make the experiment and to ask the question whether this "given" would not be sufficient for also understanding on the basis of this kind of thing the so-called mechanistic (or "material") world? I mean, not as a deception, as "mere appearance," an "idea" (in the sense of Berkeley and Schopenhauer) but as holding the same rank of reality as our affect—as a more primitive form of the world of affects in which everything still lies contained in a powerful unity before it undergoes ramifications and developments in the organic process (and, as is only fair, also becomes tenderer and weaker)—as a kind of instinctive life in which all organic functions are still synthetically intertwined along with self-regulation, assimilation, nourishment, excretion, and metabolism—as a pre-form of life.

Nietzsche argues this (thought) "experiment" is demanded by a "moral of method":

> Not to assume several kinds of causality until the experiment of making do with a single one has been pushed to its utmost limit (to the point of nonsense, if I may say so)…; suppose all organic functions could be traced back to this will to power and one could also find in it the solution of the problem of procreation and nourishment—it is one problem—then one would have gained the right to determine all efficient force univocally as—will to power.

The argument begins by assuming Boscovich's model for the composition of matter can be used to accurately describe the activity of the soul, and it concludes that we have good reason to consider whether this model can be used to describe all the phenomena in the material world. I will refer to this as Nietzsche's *will to power argument*.

The philosopher R. Lanier Anderson describes how the will to power argument is an argument for the unity of science. "The moral of method" Nietzsche mentions refers to a preference for an "economy of principles" (*BGE* 13). Anderson notes that if the OWP provides a better explanation for natural phenomena than other alternative theories in some discipline of science, this will create a pressure on other disciplines to consider whether the OWP also provides a better explanation for the phenomena investigated (2005, p. 82). This is what motivates Nietzsche's attempt to apply Boscovich's model for matter to the activity of the soul and all natural systems. Nietzsche does not argue for a reduction of all scientific disciplines to thermodynamics; he argues for a unified view of science, where disciplines work together to provide the most accurate descriptions of phenomena using an economy of principles.

In his will to power argument, Nietzsche describes himself outlining an "experiment" (*BGE* 36). The OWP succeeds or fails based on its ability to explain natural phenomena better than the available alternative theories. Like any other scientific theory, the OWP can be replaced if a new theory emerges that provides a better explanation of natural phenomena. The OWP is therefore provisional in the sense required by Nietzsche's perspectivism (Anderson, 2005, p. 78). Many interpreters have acknowledged that Nietzsche did not describe the OWP as an *a priori* metaphysical truth; he describes it as an empirical hypothesis that should be tested, like any other scientific hypothesis.[8]

My investigation is focused on those aspects of the will to power that are consistent with developments in the contemporary sciences, specifically the MPP; it is not intended to provide a comprehensive analysis of all the influential interpretations of the will to power. Consequently, a number of valuable interpretations of the will to power have not been discussed.[9] The different aspects of the OWP discussed here are all consistent with the MPP: the MPP is an empirical principle; it describes natural systems evolving in a manner that increases their power; it applies to abiotic and biotic systems, as well as human social systems; it entails the three senses of transformation outlined; and it views all systems as being products of the interaction among the parts of the system within particular environments. Consequently, the empirical evidence and the theoretical support for the MPP discussed in the last chapter also provide support for the OWP.

5.2.3   Influential Interpretations of the Will to Power

The will to power has inspired much debate about its nature and the role it plays in Nietzsche's philosophy. It has been referred to as one of Nietzsche's most controversial concepts (Emden, 2014, p. 30). Interestingly, it has come to be a part of Nietzsche's work upon which philosophers have forged a relative level of agreement: most philosophers view the OWP to be empirically implausible according to the sciences in Nietzsche's time as well as the contemporary sciences. As such, this concept is viewed as being inconsistent with Nietzsche's commitment to the value of scientific inquiry, which is referred to in the secondary literature as his *naturalism*. Christopher Janaway suggests that Nietzsche's concept of the will to power provides problems for naturalistic interpretations of his work because it is offered as a critique of the sciences in his time (2007, p. 52). Brian Leiter suggests that the

---

[8] See Richardson (1996, pp. 178–180), Emden (2014, ch. 5), Anderson (2005, p. 78), Helmut Heit (2016), and Stephen P. Schwartz (1993). In the wake of our discussion of Odum's systems ecology, one might see the hint of a systems approach to Nietzsche's argument for the will to power: the mind and matter are both viewed as systems that are the products of interacting parts and the conclusion is that everything in nature can be explained using this model.

[9] See, for example, Gunter Abel (1984), Wolfgang Müller-Lauter (1999), and Mattia Riccardi (2014). Müller-Lauter's discussion of Nietzsche's view that nature is based on antagonistic relations of forces is certainly consistent with the view of the will to power developed in this investigation.

OWP provides a challenge to naturalistic interpretations of Nietzsche's philosophy because it is based on a "crackpot" view of metaphysics (2013, p. 594).

The concerns about the OWP led many interpreters to argue that the will to power really applies only to organic systems. John Richardson provides one of the most detailed examinations of the will to power in his book *Nietzsche's System* (1996). Richardson writes that the OWP represents Nietzsche's "dominant view" because it "pulls together the greatest share of his other main ideas" (2002, p. 538); however, in his next book on Nietzsche, he decided to focus on the will to power understood as a biological principle because the metaphysics outlined by the OWP "has small plausibility for most of us" (2004, p. 12).

The concerns about the OWP have led some interpreters to argue that the will to power really describes only psychological phenomena. Maudemarie Clark describes her own psychological interpretation of the will to power as being motivated by a desire to naturalize Nietzsche's concept: to interpret it in a way that is fully supported by the contemporary sciences and therefore more consistent with Nietzsche's naturalism (2000, p. 120).[10] She suggests that this makes her interpretation more defensible than other more traditional interpretations. She along with David Dudrick developed a decidedly untraditional, "esoteric" interpretation of Nietzsche's work that goes to great lengths to suggest that Nietzsche did not intend for his readers to conclude that he accepted the OWP, even though he offers an argument for it (*BGE* 36): they argue that he did not mean what he appears to have written (2012). In the process, they engage in stretches and leaps in their interpretation that have been described as "remarkable" (Schacht, 2014, p. 352). Clark and Dudrick actually suggest that the OWP raises questions about "Nietzsche's status as a serious philosopher" (2012, p. 138). The assumption that the OWP is empirically implausible has had a profound impact on the secondary literature.

This book makes the case that we do not have to engage in any spectacular, esoteric interpretations in order to describe Nietzsche's work as being consistent with the contemporary sciences. The OWP is not only empirically plausible, but 20th and twenty-first century scientists in the "thermodynamic school" of evolution have used a version of it that is actually called the *maximum power principle*. None of the philosophers who have criticized the OWP as being empirically implausible have mentioned the MPP, the principle of maximum energy flux, or the maximum entropy production principle; they have not mentioned the work of Lotka, Odum, or the extensive literature from any of the other scientists that have worked on the maximum entropy production principle. Nietzsche's research on the OWP is informed by the work of eighteenth and nineteenth century scientists and it is supported by the work of 20th and twenty-first century scientists. Consequently, the OWP does not provide a problem for naturalistic interpretations of Nietzsche's work; it provides more support for them. A few philosophers have investigated the relation between the will to power and the work of eighteenth and nineteenth century scientists; some

---

[10] Also see, John Richardson (2000, p. 109).

of this work was discussed in Chap. 2.[11] Philosophers have also described how many of Nietzsche's ideas have come to be supported by contemporary scientific developments.[12] However, philosophers have generally not turned to the work of 20th and twenty-first century scientists to see if it provides empirical support for the OWP.

Matthew Meyer is an exception here. He recently wrote a paper which demonstrated that developments in the contemporary philosophy of science provide support for Nietzsche's dynamic relational ontology, which is a part of his OWP (2018). This book demonstrates how developments in 20th and twenty-first century science provide support for a different aspect of Nietzsche's OWP: his thesis that natural systems develop in ways that increase their power. This book, along with the work done by Meyer, demonstrate that Nietzsche's OWP is supported by the work of contemporary scientists and philosophers of science and therefore warrants consideration in contemporary discussions of philosophy.

5.3 Nietzsche's Metaethical Epistemology

Nietzsche's OWP plays a fundamental role in his critique of morality. This section will provide a textual argument for a unique interpretation of Nietzsche's critique of morality, which demonstrates that he does not assess the value of moral values from the perspective of other moral values; he assesses them from a scientific perspective: more specifically, he *describes* the affects moral values have on the growth and flourishing of life, as outlined by an empirical theory of social development based on the OWP. This is Nietzsche's view of a fully naturalized moral philosophy: what he calls a "science of morals" (*BGE* 186). This investigation will then illustrate the differences between this interpretation of Nietzsche's critique of morality and other influential interpretations.

Since the turn of the century, there have been some important efforts to develop a better understanding of Nietzsche's critique of morality by Brian Leiter (2000), Nadeem Hussain (2007), and Paul Katsafanas (2011), among others. These essays seek to provide, among other things, an answer to the fundamental questions: how does Nietzsche justify his critique of morality? Do the values he uses to assess moral values have a privileged metaphysical or epistemological status?[13] Are they privileged because they are true? Are they justified in some unique way? Are they not privileged at all? Are they simply the values Nietzsche chooses to use? This section provides an answer to these fundamental questions.

However, to make these answers clear to an interdisciplinary audience, I need to define some of the terms that will be used in this discussion. The philosophical study of ethics is usually divided into at least three areas: *normative ethics, practical*

---

[11] See Greg Moore (2002), John Richardson (2004), and Christian Emden (2014).

[12] See Brian Leiter (2013, section IV), and Leiter and Joshua Knobe (2007, ch. 4).

[13] Leiter (2000, p. 277).

ethics, and metaethics. Normative ethics focuses on the analysis of different ethical theories that provide a definition for the behavior that is considered to be moral or immoral using different criteria. One of the most famous ethical theories is *utilitarianism.* It holds that actions are moral if they maximize happiness, utility, or preference satisfaction more than the alternatives. Practical ethics applies normative ethical theories to practical issues, like abortion or euthanasia, and considers the moral status of the different forms of behavior associated with these issues. Metaethics focuses on foundational issues in the study of ethics.

For example, *metaethical semantics* addresses questions concerning the precise meaning of ethical statements. *Moral realism*, for example, is a view in metaethical semantics which holds that there are moral facts that are not dependent on human minds and humans can make claims about them that are true or false. *Anti-realism* is a view that rejects moral realism. One form of anti-realism is *non-cognitivism*; it holds that moral judgments are not propositions that can be true or false. *Constitutivism* illustrates that all actions have structural features—goals, principles, or standards—and these features can have normative (evaluative) implications for how people ought to behave that can be referenced in moral claims. *Fictionalism* holds that moral claims are fictional claims which are not true. *Expressivism* holds that moral judgments are merely expressions of subjective attitudes or feelings. From this perspective, if people say abortion is morally wrong, they are merely expressing their subjective attitudes about abortion; they are not articulating a proposition that can be considered true or false.

Metaethical epistemology focuses on issues relating to our ability to know and/ or justify moral claims: how do we justify moral claims? Can they be justified? Epistemology is a branch of philosophy that focuses on understanding the nature of knowledge. Metaethical epistemology is related to metaethical semantics in many respects; for example, fictionalism would imply that moral claims cannot be justified. There are three major positions in metaethical epistemology: empiricism, moral rationalism, and moral skepticism. *Empiricism* holds that knowledge is gained primarily through experience and observation, and this includes knowledge about moral values. *Moral rationalism* holds that moral truths are known *a priori*, by reason alone, not through experience. *Moral skepticism* denies or raises doubts about moral facts or properties and the ability to justify moral claims.

Ethical naturalism is a view that holds that moral issues can be investigated using scientific methods. These days, it tends to be construed as a form of moral realism (see, for example, Nuccetelli et al., 2012, 1). There have been increasing doubts over the past few decades about the objections to realism raised by arguments such as G. E. Moore's open question argument (which will be discussed later), and there have also been key developments in the philosophy of mind and language (Nuccetelli et al., 2012, 1), which have both led to a renewal of interest in moral realism. However, new concerns have also been raised about the relation between moral realism and the sciences. Sharon Street has argued compellingly that evolution has had a profound influence on our evaluative attitudes, and that realist interpretations of moral value are unable to provide a plausible explanation for this influence (2006). This investigation demonstrates that Nietzsche outlines a unique, comprehensive, antirealist version of ethical naturalism that can make a provocative contribution to the contemporary debate.

5.3.1 The Will to Power as an Empirical Principle

If some value has a privileged normative status, it entails that some behavior or mental state is justified. If some proposition has a privileged epistemological status, it entails that our belief in it is justified. My argument begins by accepting a view adopted by Hussain and Leiter which holds that Nietzsche does not describe the will to power as having a privileged *normative* status, but it rejects their conclusion that Nietzsche describes it having no privileged *epistemological* status. We begin to understand the significance of this difference when we acknowledge the fact that Nietzsche describes the will to power as an empirical principle, not a moral value. As illustrated before, Nietzsche argues, *"physiologists"* and *"natural scientist[s]"* should recognize the "will to power" is "the will of life" (*BGE* 13; *GS* 349. My emphasis). He describes life growing "not out of any morality or immorality, but because it is alive, and because life is precisely will to power" (*BGE* 259). In these passages, it is clear that he does not view the will to power as a fundamental moral value; he describes it as the empirical principle that guides the evolutionary development of natural systems (*BGE* 36).[14] He, therefore, argues the OWP has a privileged epistemological status because it accurately describes evolutionary development.

5.3.2 The Will to Power and Social Growth

I argue here that Nietzsche takes a more radical, naturalistic turn in his critique of morality that is overlooked by philosophers in the secondary literature. Nietzsche uses the will to power as a foundation for an empirical theory of social development that provides a thermodynamic framework for the growth and flourishing of social systems. He sees societies that are more powerful having a selective advantage in evolutionary processes. He describes the effects moral values have on the growth of social systems from the perspective of this framework. The model of growth described by this framework comes into relief when it is viewed over long periods of time. Nietzsche implicitly describes the effect the Neolithic revolution had on the

---

[14] Richardson describes the will to power as a biological principle in *Nietzsche's New Darwinism* (2004). David Owen (2003) and Christian Emden (2016). Richardson describes the will to power as an ontological principle in *Nietzsche's System* (1996); Joseph Keeping describes the will to power as an empirical principle but he utilizes the psychological interpretation of the concept in his essay "The Thousands Goals and the One Goal: Morality and the Will to Power in Nietzsche's Zarathustra" (2012).

emergence and evolution of morality (*GM* II: 16; *BGE* 32): he focuses on the development of human life on an evolutionary time scale—tens of thousands of years. The kind of growth Nietzsche describes with the will to power is the long-term growth we see reflected in the evolution of the human modes of production: hunting and gathering, agricultural production and the domestication of animals, industrial production, and information-based production. Each step of this evolution is marked by a non-linear increase in the power of the production process. The growth of this specific form of power is clearly not the only thing Nietzsche refers to when he applies the will to power to human social systems. These social systems manifest many different forms of power. However, the growth in the power of the process of production is clearly one of the fundamental parts of the growth of social systems that is described by the will to power.

We see a glimpse of this long-term tendency to grow in human social systems in the passage below. Behind all the moral and political foregrounds of Europe's democratic movement, Nietzsche writes that,

> … a tremendous *physiological* process is taking place and gaining momentum. The Europeans are becoming more similar to each other; they become more and more detached from the conditions under which races originate that are tied to some climate or class; they become increasingly independent of any determinate milieu that would like to inscribe itself for centuries in body and soul with the same demands. Thus an essentially supranational and nomadic type of man is gradually coming up, a type that possesses, physiologically speaking, a maximum of the art and power of adaptation as its typical distinction. (*BGE* 242)

Here, Nietzsche refers to the process of production as a "tremendous *physiological* process" and he explains how the growth of this process has made it possible for Europeans in his time to enjoy an unprecedented measure of freedom, relative to their predecessors who lived in agricultural societies or in hunting and gathering groups. He wrote this during the second industrial revolution (1870–1914). Like Lotka, the historian Marc Overton describes the agricultural revolution in England, or the second agricultural revolution (1650–1770), improving agricultural productivity in a manner that freed up workers to focus on different areas of the growing economy: enabling them to become "increasingly independent of any determinate milieu" (Overton, 1996).

Nietzsche describes in his notes how the "justifying man" who creates higher forms of being and new forms of power depends on the power of the productive process.

> Once we possess that common economic management of the earth that will soon be inevitable, mankind will be able to find its best meaning as a machine in the service of this economy—as a tremendous clockwork, composed of ever smaller, ever more subtly "adapted" gears; as an ever-growing superfluity of all dominating and commanding elements; as a whole of tremendous force ….
>
>   In opposition to this dwarfing and adaptation of man to a specialized utility, a reverse movement is needed—the production of a synthetic, summarizing, justifying man for whose existence this transformation of mankind into a machine is a precondition, as a base on which he can invent his *higher form of being*.
>
>   He needs the opposition of the masses, of the "leveled," a feeling of distance from them! He stands on them, he lives off them. (*WP* 866; Nachlass 1887, 10[17] *KSA* 12.462–463)

This "justifying man" stands on the economy that is served by the work of those adapted to specialized forms of utility: it is the "base on which he can invent his *higher form of being*."

When the productive power of a society increases, this "justifying man" becomes "increasingly independent of any determinate milieu"; he has more freedom to develop higher forms of being and new forms of power (*BGE* 242). When we combine the point Nietzsche makes at *WP* 866 with the one he makes at *BGE* 242, we have a simple principle:

P1) When the power of the productive process increases, it expands the limits on a society's ability to manifest new forms of power.

With this principle and his concept of the will to power, Nietzsche outlines a general thermodynamic framework for the growth and flourishing of social systems.

A point of contention in interpreting Nietzsche is who is the recipient of this growth and flourishing? Leiter interprets Nietzsche's references to the growth and flourishing of life as applying only to certain types of individuals, the "higher men" (2000: 289). Bernard Reginster criticizes this interpretation; he argues that Nietzsche does not relativize the concept of flourishing: he refers to "human flourishing"— "the highest power and splendor actually possible to the type man (*GM* Preface 5–6)" (Reginster, 2003). The view outlined in this investigation is more consistent with Reginster's interpretation, but these two views can be understood as supporting each other: the "higher men" to whom Leiter refers are needed to lead others to the kind of human flourishing on which Reginster focuses, and these "higher men" stand on the masses that serve the growth of the economy.

This thermodynamic framework can be understood as a further development of the "scaffolding" concept Emden uses in his description of the will to power (2016, p. 50).[15] He points out that the will to power does not entail any specific ethical commitments; it creates a "space of possibilities" within which people can engage in different forms of behavior. The genetic structure of human beings, for example, eventually made it possible for them to speak; it did not force human beings to say anything in particular: it created a "space of possibilities" within which different things could be said. Emden suggests that it might be useful to see the will to power as a kind of "scaffolding" that creates a "space of possibilities" within which people can engage in different forms of behavior, all of which from a moral perspective would be value neutral. Here, the concept of "scaffolding" is generically understood as something that enables the accomplishment of an outcome, like repairing a tall building.

However, the will to power is not static; it describes how things "grow" and "spread" (*BGE* 259). The thermodynamic framework for growth described above illustrates how this "scaffolding" grows over time, expanding the "space of possibilities." We see evidence of this expansion in *BGE* 242. In Nietzsche's time, the space of possibilities had opened up for human beings considerably, providing

---

[15] For more information on this concept of "scaffolding," Emden cites the work of William C. Wimsatt and others (Wimsatt et al., 2007; Wimsatt, 2001; and Caporeal et al., 2013).

people the ability to live in different environments and make a living in ways they could not a couple thousand years earlier, *e.g.*, as a train engineer or a telegraph operator. When we acknowledge that the scaffolding grows over time, it opens the way to a new interpretation of Nietzsche's critique of morality: instead of viewing him as criticizing moral values from the perspective of other moral values, we can view him as *describing* the effects moral values have on the growth and flourishing of life—understood as the growth of the scaffolding and the expansion of the space of possibilities. If Nietzsche was not aware that this scaffolding grows over time, he would not be in a position to describe moral values fostering or undermining its growth.

Nietzsche, for example, describes Christianity as placing "the emphasis of life ... on the 'beyond' rather than on life itself" (*A* 43). In the process, "the emphasis has been completely removed from life." As a result, "everything beneficial and life-enhancing in the instincts, everything that guarantees the future, now arouses mistrust" (*A* 43): "What is the point of public spirit, of being grateful for your lineage or for your ancestors, what is the point of working together, of confidence, of working towards any sort of common goal or even keeping one in mind?" When a value system describes an imaginary life in heaven being more valuable than the life we experience, one can reasonably conclude that it will undermine the ability of people to foster the growth and flourishing of the life we experience.

Arguing that a value system is immoral is another matter; Nietzsche recognizes this. He does not criticize Christianity for being immoral; he *describes* it undermining the growth of life. He argues that it takes the part of "all the weak, the low, the botched" and it makes "an ideal out of *antagonism* to all the self-preservative instincts of sound life..." (*A* 5). Nietzsche makes it clear this is *not* a "moral accusation," but just a description. He writes,

> It is a painful, horrible spectacle that has dawned on me: I have drawn back the curtain from the *corruption* of man. In my mouth, this word is at least free from one suspicion: that it might involve a moral accusation of man. It is meant—let me emphasize this once more— *moraline-free* [*moralinfrei*], so much so that I experience this corruption most strongly precisely where men have so far aspired most deliberately to 'virtue' and 'godliness.' I understand corruption, as you will guess, in the sense of decadence: it is my contention that all the values in which mankind now sums up its supreme desiderata are *decadence-values*. (*A* 6)

Nietzsche describes Christian values as being decadent. He uses the terms *decadent* and *decadence* frequently in the books he published after *BGE*.[16] He writes, "Wherever the will to power declines in any form, there is invariably also a physiological retrogression, decadence." (*A* 17). He uses the term "decadence" to refer to those things that undermine the growth and flourishing of life as outlined by the will to power.

Nietzsche uses other terms to refer to things that foster the growth and flourishing of life. Hussain explains that throughout his writings, Nietzsche uses a number

---

[16]GM – 36, A – 36, TI – 30, EH – 36, CW – 27.

of different terms to implicitly refer to the will to power: a strong will, health, creativity, vigor, self-reliance, and intelligence, among others (2013, p. 400). To be consistent with the point Nietzsche makes in the passage above (*A* 6), his use of these "evaluative" terms should not be characterized as moral evaluations, but as *descriptions* of things that foster the growth and flourishing of life. This interpretation of Nietzsche's critique of morality is consistent with his description of his own project as translating "humanity back into nature," leading us to "stand before the human being, just as he already stands before the rest of nature today, hardened by the discipline of science" (*BGE* 230).

5.3.3 Nietzsche's "Science of Morals"

In *Beyond Good and Evil* (*BGE* 186), Nietzsche discusses the problems he sees with the "science of morals" that has been practiced in Europe in his time. This science, practiced, for example, by the German physician and philosopher Paul Ree, Nietzsche's one time friend, posits that moral issues can be investigated in an inductive manner (Ree, 2003). Nietzsche's comments on this topic provide some of the clearest textual evidence that he views himself as describing the effects moral values have on the growth and flourishing of life from a scientific perspective. He begins by describing this "science of morals" in his time as being "still young, raw, clumsy, and butterfingered." He suggests that it has so far lacked two things: "the problem of morality itself" has not been acknowledged, and a "typology of morals" has not been developed.

First, Nietzsche argues that those who practice this "science of morals" in his time have not considered "the problem of morality itself." They have lacked,

> any suspicion that there was something problematic here. What the philosophers called "a rational foundation for morality" and tried to supply was, seen in the right light, merely a scholarly variation of the common *faith* in the prevalent morality; a new means of *expression* for this faith; and thus just another fact within a particular morality; indeed, in the last analysis a kind of denial that this morality might ever be considered problematic—certainly the very opposite of an examination, analysis, questioning, and vivisection of this very faith. (*BGE* 186)

Instead of critically analyzing this faith, "every philosopher so far has believed that he has provided such a foundation. Morality itself, however, was accepted as 'given.'"

Early in *BGE*, Nietzsche describes the "stiff and decorous Tartuffery of the old Kant as he lures us on the dialectical bypaths that lead to his 'categorical imperative'—really lead astray and seduce—this spectacle makes us smile, as we are fastidious and find it quite amusing to watch closely the subtle tricks of old moralists and preachers of morals" (*BGE* 5). Nietzsche asks us to consider "the hocus-pocus of mathematical form with which Spinoza clad his philosophy." He concludes that philosophers, all pose as if they had discovered and reached their real opinions through the self-development of a cold, pure, divinely unconcerned dialectic (as opposed to the mystics of every rank, who are more honest and doltish—and talk of "inspiration"); while at bottom it is an assumption, a hunch, indeed a kind of "inspiration"—most often a desire of the heart that has been filtered and made abstract—that they defend with reasons they have sought after the fact. They are all advocates who resent that name, and for the most part even wily spokesmen for their prejudices which they baptize "truths...." (*BGE* 5)

Thus, in Nietzsche's view, all philosophers pose as if they have discovered a "rational foundation for morality" when, in actuality, they have merely offered "a new means of expression" for the "faith" in morality.

A few years later in 1888, Nietzsche describes in his notes the critique of morality outlined by the ancient Sophists. Rather than citing the failures of moral philosophers throughout history in a diachronic fashion, they focus on the failures of moral philosophers across different cultures during the same time, in a synchronic fashion. He writes:

It is a very remarkable moment: the Sophists verge upon the first critique of morality [*Moral*], the first insight into morality:—they juxtapose the multiplicity (the geographical relativity) of the moral value judgments [*Moralischen Werthurtheile*];—they let it be known that every morality [*Moral*] can be dialectically justified; i.e., they divine that all attempts to give reasons for morality [*Moral*] are necessarily sophistical—a proposition later proved on the grand scale by the ancient philosophers, from Plato onwards (down to Kant);—they postulate the first truth that a "morality-in-itself" [*eine Moral an sich*], a "good-in-itself" do not exist, that it is a swindle to talk of "truth" in this field. (*WP* 428; Nachlass 1888, 14[116], *KSA* 13.292–293).

Thus, Nietzsche describes the Sophists as being the first to provide an argument that leads us to question our "faith" in morality and consider "the problem of morality itself."

In his discussion of the "science of morals" in his time, Nietzsche refers to the will to power, but he does not describe it as a "rational foundation for morality." He suggests Schopenhauer's fundamental principle of morality—"*neminem laede, immo omnes, quantum potes, juva* [Hurt no one; rather, help all as much as you can]....*"—can appear "insipidly false and sentimental" in a "world whose essence is will to power" (*BGE* 186). He does not refer to the will to power as a fundamental moral value; he refers to it as an empirical principle, as he does throughout the book (*BGE* 13, 23, 36, 259). The whole passage makes it clear he believes a "science of morals" needs to question the assumption that there is a "rational foundation for morality," and it implicitly describes his own "science of morals" as being based on an empirical foundation: the will to power.

In Nietzsche's discussion of "the problem of morality itself," we see an argument for why we should evaluate moral values from the empirical perspective outlined by the will to power. He describes how philosophers have continually tried and failed to provide a "rational foundation for morality," but this lack of success has not led them to question whether morality may not have a rational foundation: they have not considered the "problem of morality itself." He suggests their pursuit of a rational foundation has become an expression of a kind of "faith" that is not supported by the evidence: it is "certainly the very opposite of an examination, analysis, questioning, and vivisection of this very faith." Nietzsche implies here that an "examination" or "vivisection" of this "faith" is called for given the evidence and the nature of the issue.

The argument he makes, which is to a certain extent implicit, exhibits a form of scientific rationality: the fact that philosophers so far have failed to provide a "rational foundation for morality" cannot be used to demonstrate that it is not possible; however, it can be used as empirical evidence that justifies an investigation of whether a science of morals that has an empirical foundation can provide a better understanding of what morality is and how it functions. Nietzsche describes the history of moral philosophy providing empirical evidence, both diachronic and synchronic, that supports his own "science of morals." I will refer to this argument as Nietzsche's *"problem of morality itself"* argument.

We can see Nietzsche leaning on the problem of morality itself argument and the will to power argument later in *BGE*. In *BGE* 230, Nietzsche describes all living things having an "intent" to grow and he describes human beings "hardened in the discipline of science" with "intrepid Oedipus eyes and sealed Odysseus ears, deaf to the siren songs of old metaphysical bird catchers who have been piping at him all too long, 'you are more, you are higher, you are of a different origin!'" Here, he is clearly trying to persuade us that it does not make sense for humans to act as if we have a different "intent" than all other living things. His alternatives are, however, limited. He cannot offer a valid argument that it is immoral to do so because he does not think life grows because it is immoral not to (*BGE* 259). What Nietzsche can do is, first, highlight the empirical evidence that there is a tendency in all natural systems to grow, as he does in the will to power argument (Sect. 5.2.2); and, second, he can remind us that for over two thousand years the "experts" have not been able to provide a "rational foundation" for any moral theory which holds that humans ought to strive for something other than the growth and flourishing of life, as he does in the problem of morality itself argument. The point Nietzsche makes at *BGE* 230 is therefore supported by the two inductive or abductive arguments he makes earlier in *BGE*: the will to power argument (36) and the problem of morality itself argument (186).

The thesis that the problem of morality itself argument concludes that a science of morals based on an empirical foundation should be investigated is further supported by Nietzsche's discussion of the need for "a typology of morals." The second thing Nietzsche believes the "science of morals" in his time lacks is a "typology of morals."

> One should own up in all strictness to what is still necessary here for a long time to come, to what alone is justified so far: to collect material, to conceptualize and arrange a vast realm of subtle feelings of value and differences of value which are alive, grow, beget, and perish—and perhaps attempts to present vividly some of the more frequent and recurring forms of such living crystallizations—all to prepare a typology of morals.

He characterizes this work as being based on "the task of description," a task philosophers have viewed as "insignificant and left in dust and must...although the subtlest fingers and senses can scarcely be subtle enough for it." Because philosophers have overlooked this task, "they never laid eyes on the real problems of morality; for these emerge only when we compare many moralities." Nietzsche begins to bring these "real problems" into relief in his next book *On the Genealogy of Morals* (1887), where he starts to develop the typology (or taxonomy) of morals described above.[17]

When we acknowledge that Nietzsche views the will to power as an empirical principle and he views himself as *describing* the effects moral values have on the growth and flourishing of life from a scientific perspective, we can begin to see that Nietzsche's critique of morality addresses the two things he describes the "science of morals" in his time lacking: it critically examines the "problem of morality itself;" and it begins to develop a "typology of morals." Nietzsche's critical analysis of the "science of morals" in his time provides a revealing look at his own view of his critique of morality: he sees it as a "science of morals" that is no longer "young, raw, clumsy, and butterfingered."

The thesis that Nietzsche views himself as assessing moral values from a scientific perspective is also supported by his description of himself as a "physician" of culture who reads moral values as "a group of symptoms" and knows how to "interpret them correctly" in order to "profit from them" (*A* 7; *TI* "Improvers" 1). He writes that, "moral judgments … reveal …the most valuable realities of cultures" and psychologies (*TI* "Improvers" 1). The physician of culture understands these realities and cuts away that which is decadent and prescribes "natural" or "healthy" moralities that remove "hostile" elements "on the path of life" (*TI* "Anti-Nature" 4). Nietzsche adopts what Daniel Ahern has called a "clinical point of view" (1995) and describes his philosophy as a kind of therapy, similar to the therapeutic approach used by the Epicureans and the Stoics.[18]

We have evidence that people have followed the advice of physicians on health issues since the time of Hippocrates of Kos (460 BC–370 BC). As the science of medicine has grown, along with other scientific disciplines, the "health issues" for which the sciences have provided guidance have increased; now, they not only provide guidance on how to treat illnesses, but also on how to avoid illnesses, what to eat, how to exercise, deal with emotional problems, and live a long life; in the contemporary sustainability movement, science also provides guidance on how to foster the long-term growth and flourishing of life.[19] The extent to which science has influenced human behavior has increased along with the growth of science itself. Nietzsche's "science of morals" is consistent with this trend, even though it also represents a radical departure from traditional approaches to moral philosophy. This should   mitigate   any   concerns   about   why   someone   would   follow   the

---

[17] This point is made by Timothy Soll (2018, p. 546).

[18] Also see Marta Faustino (2017) and Leiter (2019, p. 6).

[19] The *Leadership in Energy and Environmental Design* (LEED) rating system has become very influential in the construction industry. It seeks to encourage the use of sustainable practices. Universities and Colleges around the country use this rating system to guide their decision. See the United States Green Building Council's LEED program: https://new.usgbc.org/leed

recommendations of Nietzsche's "science of morals," i.e., what philosophers have called its *normative oomph.*

5.3.4 Influential Interpretations of Nietzsche's Metaethics

This section critically analyzes some of the most influential interpretations of Nietzsche's metaethics. In the secondary literature, we find that philosophers generally agree that Nietzsche reevaluates moral values from the perspective of the will to power, but they disagree about the status of the will to power. Some interpreters argue that the will to power is a fundamental moral value that has a privileged normative status: Richardson, Katsafanas, Schacht, and Wilcox. This view has been interpreted as a form of *moral realism.* Katsafanas views the will to power as a fundamental moral value, but he justifies it using a form of *constitutivism.* Other interpreters argue that the will to power has no privileged normative or epistemological status. Nadeem Hussain initially described Nietzsche's view of the values he uses to guide his critique of morality as a form of *fictionalism*, which are not privileged in any way. He eventually reconsidered this position, but he continues to agree with Brian Leiter that the will to power does not have a privileged normative or epistemological status: Nietzsche's views on moral values are simply the positions he freely chooses to take. Leiter describes Nietzsche's position as a form of *moral skepticism.*

We will begin by critically analyzing the view that the will to power is a fundamental moral value that has a privileged normative status. This view has become a kind of "orthodoxy" and a metaethical view (Hussain, 2007, p. 176; Richardson, 1996, chapter 3). The philosopher Walter Kaufmann believed that for Nietzsche the measure of value was the quantity of power (1974, p. 200). There is textual evidence for this interpretation: Nietzsche writes, "What is good? Everything that enhances people's feeling of power, will to power, power itself. What is bad? Everything stemming from weakness" (*A* 2). Hussain argues that this interpretation of the will to power commits Nietzsche to a form of *moral realism*: to be powerful is what it means to be good (2007, p. 177). Leiter describes Richard Schacht and John Wilcox defending a similar interpretation of Nietzsche's metaethical position, which he refers to as a *natural realism* (2000, p. 277).

Katsafanas outlines a unique way to interpret the will to power as the ultimate standard of moral value (2011). He argues that Nietzsche uses the constitutive features of action to derive a standard of success for all actions that justifies a privileged normative status for the will to power. The will to power is a fundamental tendency to overcome resistance in order to achieve an objective. This tendency is fundamental to most forms of action and, therefore, it has a particular normative significance. Nietzsche's strategy therefore has parallels to that used by contemporary constitutivists. Katsafanas notes that constitutivism has historical roots in the work of Plato, Aristotle, and Kant, philosophers with whom Nietzsche was deeply engaged, so it would not be surprising if Nietzsche developed a theory of value that took the form of constitutivism.

A primary problem with any attempt to describe the will to power as a fundamental moral value is that, as Hussain argues, it characterizes the will to power as something that is valuable in itself, notwithstanding the fact that Nietzsche wrote that nothing has value in itself (2007, p. 177).[20] Nietzsche is clear: *"there are absolutely no moral facts"* (*TI* "Improvers" 1). He encourages us to "stand beyond good and evil and treat the illusion of moral judgment as beneath" us. Arthur Danto acknowledges that, according to Nietzsche, nothing in the world has value (1965, p. 33).[21] Hussain therefore concludes along with Leiter that Nietzsche does not describe the will to power as having a privileged normative status. Hussain argues in the end that Nietzsche's texts lack the specificity needed to resolve the issues associated with the different contemporary metaethical positions (2013, p. 412), whether it is the fictionalism he initially advocated, the moral realism described by Richardson and Schacht, or the noncognitivism described by Maudemarie Clark and David Dudrick. Hussain and Leiter both argue that Nietzsche did not believe the will to power had any privileged metaphysical, epistemic, or normative status (Leiter, 2000, p. 290). They both believe that the textual evidence suggests that the will to power is simply the perspective Nietzsche freely and arbitrarily chooses to take.

Hussain and Leiter and other philosophers have looked in vain in Nietzsche's texts for a valid argument for the will to power as moral value. Hussain considers what he calls a *"Benthamite"* argument. He notes that Bentham believed "nature has placed mankind under the governance of two sovereign masters, pain and pleasure.… The principle of utility recognizes this subjection" (Bentham, 2003, p. 17). Hussain suggests Nietzsche appears to be making a similar argument with respect to the will to power: all life strives for power; Nietzsche's philosophy merely recognizes this subjection.

Hussain provides an example of a similar argument that was made in Nietzsche's time. He quotes Karl Kautsky: "[It is] the materialist conception of history which has first completely deposed the moral ideal as the directing factor of social evolution, and has taught us to deduce our social aims solely from the knowledge of the material foundations" (Lukes, 1985, p. 18). Marxists saw themselves as criticizing morality from the perspective of a scientific theory of social development based on the material conditions of production in social systems. After laying out this *Benthamite* argument, Hussain concludes that it is clearly not valid: it does not justify a belief that power has a privileged normative status. But it does appear to be the kind of argument that Nietzsche is providing for the will to power, and this conclusion is supported by the fact that others during the same historical period offered

---

[20] Ian Dunkle demonstrates other problems with Katsafanas' argument for constitutivism (2020).

[21] Hussain notes that this view is not the orthodoxy in the Anglo-American secondary literature (2007, p. 157, n. 1).

similar arguments (2011, p. 162–163). Leiter suggests that Schact describes Nietzsche offering a similar argument, which Leiter refers to as a "Millian Model," and Leiter also finds this argument to be invalid (2000, pp. 281–287).

My own investigation seeks to demonstrate that Hussain, Leiter, and other philosophers misinterpret Nietzsche's argument for the will to power as an invalid argument for the will to power as a moral value, when it is actually an inductive or abductive argument for the will to power as an empirical principle. When the will to power is understood as an empirical principle, its support is not dependent, as Hussain suggests, on a valid argument that the will to power is a fundamental moral value. Hussain's and Leiter's search for such an argument demonstrates that they have overlooked the more radical turn Nietzsche takes in his naturalistic critique of morality: he "deposed the moral ideal as the directing factor of social evolution, and has taught us to" understand our social aims from the perspective of a science of morals based on the will to power.

The argument I refer to as the problem of morality itself argument is described by Leiter as a unique form of an *argument from disagreement,* and he suggests that Nietzsche uses it to support a form of moral skepticism (2014). Leiter explains that this argument from disagreement is not focused on the many ways that people disagree on moral questions; it is focused on the fact that experts in the field of moral theory have always had intractable disagreements about the fundamental principles of morality. This persistent disagreement distinguishes moral theory from what we find in the natural sciences or mathematics, at least in degree.

Leiter suggests that if the fact that people make moral judgements was best explained by the existence of objective moral facts, first, the experts in the field of moral theory would be able to reach some sort of agreement about these facts; second, these objective moral facts would also be used in other related disciplines like sociology which is not the case. Leiter interprets this argument from disagreement as demonstrating that moral judgements are not best explained by appealing to objective moral facts; they are best explained by psychological and sociological factors that cause agents to make moral judgements. Leiter therefore views Nietzsche's argument as supporting an anti-realist form of moral skepticism.

Leiter does however note that there is at least one challenge to his interpretation. He acknowledges that Nietzsche writes "as if there really is a fact of the matter" about moral judgments (2014, p. 145). Leiter attempts to explain this tone as just being a rhetorical pose. This investigation, however, demonstrates that there is a better interpretation of Nietzsche's argument from disagreement that gives us reason to view his critique of morality as being based on something more than a rhetorical pose: his argument provides evidence that supports a science of morals that is based on an empirical foundation—the will to power; in this science of morals, he describes the effects moral values have on the growth and flourishing of life as outlined by the OWP, just like a doctor describes the effects cancer has on the growth of a teenager. There is in both cases a fact of the matter that is being described from a scientific perspective.

My thesis is that Nietzsche uses the problem of morality itself argument and the will to power argument to support his own "science of morals," which does not argue that moral values are moral or immoral; it *describes* the affects moral values have on the growth and flourishing of life, as outlined by the will to power. Those moral values and behaviors that undermine the growth of life are described as decadent, impoverished, sick, weary, discouraged, or exhausted; those that foster the growth of life are described as healthy, natural, vigorous, forceful, intelligent, and strong willed, among other things (Hussain, 2013, p. 400). This distinction is the primary focus of his critique of morality: are moral judgments "a sign of distress, of impoverishment, of the degeneration of life? Or is there revealed in them, on the contrary, the plenitude, force, and will of life, its courage, certainty, future?" (*GM* Preface 3, 6) Some of the most significant examples of textual evidence that support my thesis are listed below.

1. Nietzsche argues that the "science of morals" in his time needs to consider the "problem of morality itself" and develop a "typology of morals" (*BGE* 186).
2. He states there are no moral facts (*TI* "Improvers" 1).
3. In his will to power argument, he offers an inductive or abductive argument for the OWP as an empirical principle, not a fundamental moral value (*BGE* 36).
4. In his problem of morality itself argument, he argues that the fact that philosophers throughout the history of moral philosophy have been unable to provide a rational foundation for morality provides a justification for considering the value of a "science of morals" that is based on an empirical foundation—the OWP (*BGE* 186).
5. He goes on in *On the Genealogy of Morality*, and his other works, to *describe* the effects moral values have on the growth and flourishing of life as outlined by the OWP, i.e., he develops a "typology of morals."
6. He describes these effects "as if there really is a fact of the matter."
7. He states that his descriptions of these effects are not moral accusations (*A* 6).

In Nietzsche's vernacular, those philosophers that argue the will to power is a fundamental moral value—Richardson, Katsafanas, Schacht, and Wilcox, among others—describe the concept as being another "rational foundation for morality" and therefore they implicitly describe Nietzsche as not addressing the "problem of morality itself." The philosophers that argue that the will to power is not privileged in any way—Hussain and Leiter, among others—fail to acknowledge the central role the concept plays in Nietzsche's own "science of morals" as an empirical principle and are therefore unable to recognize that he "*describes*" the effects moral values have on the growth and flourishing of life from a perspective "hardened by the discipline of science" (*BGE* 230).

The interpretation of Nietzsche's critique of morality outlined in this investigation provides unique answers to the fundamental, metaethical questions considered by Leiter, Hussain, Katsafanas and others: how does Nietzsche justify his critique of morality? Do the values he uses to assess moral values have a privileged normative, metaphysical, or epistemological status? Nietzsche does not criticize moral values from the perspective of other moral values; he describes the effects moral values have on the growth and flourishing of life as outlined by a theory of social development based on the will to power. The privileged epistemological status of this theory is based on the empirical evidence and scientific arguments that support it.[22]

The interpretation developed in this investigation demonstrates that the metaethical epistemology implicit in Nietzsche's "science of morals" is best described as a form of *empiricism*. By empirically investigating and describing the effects different moral values have on the growth and flourishing of life, we can better understand which moral values can help foster this growth. This entails a fundamental reconsideration of the role moral values play in life, which suggests they function as instrumental tools within a larger system that is governed by thermodynamic laws and principles like the OWP or the MPP. I argue that this reconsideration is what Nietzsche calls for when he writes, *"the value of these values themselves must first be called in question"* (*GM* Preface 6), and it is similar to the reconsideration Odum calls for in his analysis of moral and religious values (1971, 1977, 2007).

There is also a metaethical semantic position implicit in the interpretation of Nietzsche's "science of morals" developed in this investigation that I can outline here only briefly. It would best be described as a variant of *expressivism*. Throughout his work, Nietzsche describes moral judgments as "signs" or "symptoms" of the growth and flourishing of life or its decline and degeneration.[23] He, for example, refers to Christianity as a "sign" of "the impoverishment of life" (*BT*, "Attempt at Self Criticism" 5). He thought moral judgments reveal "the most valuable realities of cultures" and psychologies (*TI* "Improvers" 1) and the intention behind these judgments is "merely a sign and symptom" of these realities (*BGE* 32). As mentioned, expressivism holds that moral judgments are expressions of the subjective attitudes of those who make them. Nietzsche views moral judgements as expressing more than just these subjective attitudes: they also express the psychological and cultural effects these attitudes have on the growth and flourishing of life (*BGE* 187; *GM* "Preface" 3, 6). By considering the effects of these attitudes, his view goes beyond the standard interpretation of expressivism. A full exposition and defense of this semantic position would take us well beyond the scope of this chapter, but this brief sketch can illustrate how the metaethical epistemology and semantics implicit in Nietzsche's "science of morals" work together and support each other.

---

[22] Nietzsche's views on science are quite intricate and nuanced. A detailed discussion of them would take us beyond the scope of this essay. To get a more complete picture, see Nadeem Hussain's "Nietzsche's Positivism" (2004), Maudemarie Clark and David Dudrick "Nietzsche's Post-Positivism" (2004), Lanier Anderson "Nietzsche's will to power as a doctrine of the unity of science" (1994), and Timothy Soll "Science and Two Kinds of Knowledge: Nietzsche, Schopenhauer, and the *Ignorabimus-Streit*" (2018).

[23] See, for example, *BT* 4; *BGE* 10, 187; *GM* "Preface" 3, III 25; *TI* "Improvers" 1, "Socrates" 2, "Skirmishes" 41, "Morality and Anti-nature" 5; *EH* "Birth of Tragedy" 2, "Destiny" 8; *A* 6, 50, 51.

5.3.5   Moore's Open Question Argument

The unique nature of Nietzsche's metaethical position enables his critique of morality to avoid being undermined by G. E. Moore's "open question" argument. Moore argued that the concept of "goodness" cannot be defined in terms of some natural property like happiness (or power) because we can tell that such a claim is not true by definition: the words used in the claim do not demonstrate its truth. We can always ask—is this claim true?—and the question makes sense in a way that it would not if the claim was that $2 + 2 = 4$: the claim is therefore always an "open question."

Moore's argument does not apply to Nietzsche's use of the will to power because Nietzsche does not view or use the will to power as a moral value; he views it and uses it as an empirical principle. He does not criticize moral values for being immoral; he describes them fostering or undermining the growth of life as outlined by the will to power. These are empirical claims. They, strictly speaking, will remain open questions like all scientific claims since empirical evidence can always be discovered in the future that leads us to revise our answers. Nietzsche's problem of morality itself argument suggests that the fact that throughout the history of moral philosophy no philosopher has discovered a "rational foundation for morality" that defined goodness in a closed fashion provides empirical evidence that justifies questioning the assumption or faith that there is a "rational foundation for morality." Moore's "open question" argument uses conceptual analysis to argue that it is not possible to define goodness in closed fashion. The "open question" argument therefore supports Nietzsche's *problem of morality itself argument*: it provides another way to demonstrate the "problem of morality itself."

5.4   Conclusion

This chapter described Nietzsche's OWP, his argument for it, and the role it plays in his critique of morality. It provides an argument that Nietzsche critically analyzes moral values from a scientific perspective: he describes the "effects" moral values have on the growth and flourishing of life as outlined by the OWP, understood as an empirical principle. His descriptions are based on an empirical theory of social development outlined by the OWP. This interpretation of Nietzsche's critique of morality is supported by a wealth of textual evidence found throughout his work (Sect. 5.3). The empirical evidence and theoretical arguments that support the OWP have been discussed in previous chapters (Chap. 3, Sects. 3.2 and 3.4; Chap. 4, Sects. 4.3 and 4.10).

Nietzsche's critique of morality takes a unique naturalistic approach that raises questions of its own and this chapter does not begin to answer all of them. This chapter does, however, demonstrate that his critique of morality is not threatened by Moore's open question argument; it is supported by it. This interpretation of his critique of morality can make a unique contribution to the contemporary debate over moral naturalism. Philip Kitcher has argued that taking different approaches to scientific problems increases the probability of solving them (1990). My interpretation of Nietzsche's critique or morality as laying the foundation for a "science of morals" is certainly different. Taking it into account can increase the probability that we are able to develop a more successful version of moral naturalism. Nietzsche does not dip his toes in the pool of a moral naturalism; he goes all in. In the process, he poses the question: why not confront the problem of morality itself? How much longer must we seek unsuccessfully to provide a rational foundation for morality before we investigate the value of a "science of morals" with an empirical foundation? In order to understand the role ethics actually plays, within the vast, unforgiving universe of which we are a part, Nietzsche argues that we must view moral philosophers as being cultural "physicians" trained in a "science or morals" that "wield the scalpel," cutting away that which threatens the growth and flourishing of life: "that is *our* part, that is *our* love of man, that is how *we* are philosophers..." (A 7).

In the next chapter, we will explore the relation between the work of Nietzsche and Odum, particularly the parallels between the MPP and OWP and their relation to moral and religious values. We will also consider the many differences in their analyses and what we can learn from them.

References

Friedrich Nietzsche's Works

A. (1954/1888). The Antichrist [Der Antichrist. Fluch auf das Christenthum]. In *The portable Nietzsche* (W. Kaufmann, Trans.). Viking Penguin.

BGE. (1989/1886). *Beyond Good and Evil* [*Jenseits von Gut und Böse. Vorspiel einer Philosophie der Zukunft*] (W. Kaufmann, Trans.). Vintage.

BT. (1967). *The Birth of Tragedy and The Case of Wagner,* [*Die Geburt der Tragodie*] (W. Kaufmann, Trans.). Random House.

CW. (1967/1872, 1888). *The Birth of Tragedy and The Case of Wagner* [*Die Geburt der Tragödie aus dem Geiste der Musik*] [*Der Fall Wagner*] (W. Kaufmann, Trans.). Vintage.

EH. (1967/1908). *Ecce Homo* [*Ecco Homo*] (W. Kaufmann, Trans.). Vintage.

GM. (1967/1887). *On the genealogy of morals* [*Zur Genealogie der Moral. Eine Streitschrift*]. (W. Kaufmann & R. J. Hollingdale, Trans). Vintage.

GS. (1974/1882). *The Gay Science* [Die fröhliche Wissenschaft]. (W. Kaufmann, Trans.). Vintage.

HAH. (1986). Human, all too human [Menschliches, Allzumenschliches: Ein Buch für freie Geister] (R. J. Hollingdale, Trans.). Cambridge University Press.

KSA. (1999). *Saemtliche Werke: Kritische Studienausgabe* (2nd ed.). Edited by Giorgio Colli and Mazzino Montinari. Walter de Gruyter.

TI. (1954/1888). Twilight of the idols: Or, how one philosophizes with a hammer [Götzen-Dämmerung. Wie man mit dem Hammer philosophiert]. In *The portable Nietzsche* (W. Kaufmann, Trans.). Viking Penguin.

WP. (1968). The will to power [Der Wille zur Macht]. (W. Kaufmann, Trans.). Vintage.

Z. (1954/1883). Thus Spoke Zarathustra: A book for all and none [Also sprach Zarathustra: Ein Buch für Alle und Keinen]. In *The portable Nietzsche* (W. Kaufmann, Trans.). Penguin Books.

Other Sources

Abel, G. (1984). *Nietzsche: Dynamik der Willen zur Macht und die ewige Wiederkehr*. De Gruyter.

Ahern, D. (1995). *Nietzsche as cultural physician*. Penn State University Press.

Anderson, R. L. (2005). Nietzsche's will to power as a Doctrine of the Unity of Science. *Angelaki, 10*, 77–93.

Benthem, J. (2003). *An introduction to the principles of morals and legislation, in utilitarianism and on liberty*. Blackwell.

Campbell, R. (1996). Can biology make ethics objective? *Biology and Philosophy, 11*, 21–31.

Caporeal, L. R., Griesemer, J. R., & Wimsatt, W. C. (Eds.). (2013). *Developing scaffolds in evolution, culture, and cognition*. MIT Press.

Casebeer, W. (2003). *Natural ethical facts: Evolution, connectionism, and moral cognition*. MIT Press.

Clark, M. (1990). *Nietzsche on truth and philosophy*. Cambridge University Press.

Clark, M. (2000). Nietzsche's doctrine of the will to power: Neither ontological or biological. *International Studies in Philosophy, 32*(3), 119–135.

Clark, M., & Dudrick, D. (2004). Nietzsche's post-positivism. *European Journal of Philosophy, 12*(3), 369–385.

Clark, M., & Dudrick, D. (2007). Nietzsche and moral objectivity: The development of Nietzsche's metaethics. In B. Leiter & N. Sinhababu (Eds.), *Nietzsche and morality* (pp. 111–132). Clarendon Press.

Clark, M., & Dudrick, D. (2012). *The soul of Nietzsche's beyond good and evil*. Cambridge University Press.

Danto, A. (1965). *Nietzsche as philosopher*. Columbia University Press.

Dennett, D. (1995). *Darwin's dangerous idea*. Simon and Schuster.

Dunkle, I. (2020). On the normativity of the will to power. *The Journal of Nietzsche Studies, 51*(2), 188–211.

Emden, Christian. 2014. Nietzsche's naturalism: Philosophy and the life sciences in the nineteenth century. .

Emden, C. (2016). Nietzsche's will to power: Biology, naturalism, and normativity. *Journal of Nietzsche Studies, 47*(1), 30–60.

Faustino, M. (2017). Nietzsche's therapy of therapy. *Nietzsche-Studien, 46*(1), 82–104.

Heit, H. (2016). Naturalizing perspectives: On the epistemology of Nietzsche's experimental naturalizations. *Nietzsche-Studien, 45*, 56–80.

Hussain, N. (2004). Nietzsche's positivism. *European Journal of Philosophy, 12*(3), 326–368.

Hussain, N. (2007). Honest illusions: Valuing for Nietzsche's free spirits. In B. Leiter & N. Sinhababu (Eds.), *Nietzsche and morality* (pp. 157–191). Clarendon Press.

Hussain, N. (2011). The role of life in the genealogy. In *Nietzsche's on the genealogy of morality: A critical guide* (pp. 142–169). Cambridge University Press. Kindle Version.

Hussain, N. (2012). Nietzsche and non-cognitivism. In S. Roberson & C. Janaway (Eds.), *Nietzsche, naturalism, & normativity* (pp. 111–132). Oxford University Press.

Hussain, N. (2013). Nietzsche's metaethical stance. In K. Gemes & J. Richardson (Eds.), *The Oxford handbook of Nietzsche* (pp. 389–412). Oxford University Press. Kindle Version.

Janaway, Christopher. 2007. Beyond selflessness: Reading Nietzsche's genealogy. .

Joyce, R. (2006). *The evolution of morality*. MIT.

Katsafanas, P. (2011). Deriving ethics from action: A Nietzschean version of Constitutivism. *Philosophy and Phenomenological Research, 83*(3), 620–660.

Kaufmann, W. (1974). *Nietzsche: Philosopher, Psychologist, Antichrist* (4th ed.). Princeton University Press.

Keeping, J. (2012). The thousands goals and the one goal: Morality and the will to power in Nietzsche's Zarathustra. *European Journal of Philosophy, 20*(S1), e73–e85.

Kitcher, P. (1990). The division of cognitive labor. *Journal of Philosophy, 87*, 5.

Leiter, B. (1997). Nietzsche and the morality critics. *Ethics, 107*, 250–285.

Leiter, B. (2000). Nietzsche's Metaethics: Against privilege readings. *European Journal of Philosophy, 8*(3), 277–297.

Leiter, B. (2002). *Nietzsche on morality*. Routledge.

Leiter, B. (2019). *Moral psychology with Nietzsche*. Oxford University Press.

Leiter, B., & Knobe, J. (2007). The case for Nietzschean moral psychology. In B. Leiter & N. Sinhababu (Eds.), *Nietzsche and morality*.

Leiter, B. (2013). Nietzsche's naturalism reconsidered. In J. Richardson & K. Gemes (Eds.), *The Oxford handbook of Nietzsche* (pp. 576–598). Oxford University Press. Kindle version.

Leiter, B. (2014). Moral skepticism and moral disagreement in Nietzsche. In R. Shafer-Landau (Ed.), *Oxford studies in metaethics* (Vol. 9). Oxford University Press.

Lukes, S. (1985). *Marxism and morality*. Oxford University Press.

Meyer, M. (2018). Nietzsche's ontic structural realism. In P. Katsafanas (Ed.), *The Nietzschean mind* (pp. 365–380).

Moore, G. (2002). *Nietzsche, biology and metaphor*. Cambridge.

Moore, G. (2004). Introduction. In G. Moore & T. Brobjer (Eds.), *Nietzsche and science*. Ashgate.

Müller-Lauter, W. (1999). *Nietzsche. His philosophy of contradictions and the Contradictions of His Philosophy*, translated from German by David J. Parent. University of Illinois Press.

Nuccetelli, S., & Seay, G. (2012). Introduction. In S. Nuccetelli & G. Seay (Eds.), *Ethical naturalism*. Cambridge University Press.

Owen, D. (2003). Nietzsche, revaluation and the turn to genealogy. *European Journal of Philosophy, 1*(3), 249–272.

Overton, M. (1996). *Agricultural revolution in England: The transformation of the agrarian economy 1500-1850*. Cambridge University Press.

Parfit, D. (2011). *On what matters*. Edited by S. Scheffler. Oxford University Press.

Plank, W. (1998). *The quantum Nietzsche: The will to power and the nature of dissipative systems*. Writers Club Press.

Prigogine, I. (1961). *Thermodynamics of irreversible processes*. 2nd edn.

Ree, P. (2003). *The origin of the moral sensations. Edited by Robin Small*. University of Illinois Press.

Reginster, B. 2003. Review of Brian Leiter's. *Routledge Philosophy Guidebook to Nietzsche on Morality* (London: Routledge, 2002). In *Notre Dame Philosophical Reviews*. Retrieved February 13, 2013, from http://ndpr.nd.edu/news/23223-routledge-philosophy-guidebook-to-nietzsche-on-morality/

Reginster, B. (2006). *The affirmation of life: Nietzsche on overcoming nihilism*. Oxford University Press.

Riccardi, M. (2014). Nietzsche und die Erkenntnistheorie und Metaphysik: Indizien für die konservative Lesart. In H. Heit & L. Heller (Eds.), *Handbuch Nietzsche und die Wissenschaften: Natur-, geistes- und sozialwissenschaftliche Kontexte* (pp. 242–264). De Gruyter.

Richards, R. (1986). A defense of evolutionary ethics. *Biology and Philosophy, 1*, 265–293.

Richardson, J. (1996). *Nietzsche's system*. Oxford University Press. Kindle version.

Richardson, J. (2000). Clark on will to power. *International Studies in Philosophy, 32*(3), 107–117.

Richardson, J. (2002). Nietzsche contra Darwin. *Philosophy and Phenomenological Research, 65*(3), 537–575.

Richardson, J. (2004). *Nietzsche's new Darwinism*. Oxford University Press.

Schacht, R. (1983). *Nietzsche*. Routledge.

Schacht, R. (2014). Clark and Dudrick's New Nietzsche. *Journal of the History of Philosophy, 52*(2), 339–352.

Schrödinger, E. (1945) What an organism feeds upon is negative entropy. In *What is life? The physical aspect of the living cell*. Macmillan.

Stoll, T. (2018). Science and two kinds of knowledge: Nietzsche, Schopenhauer, and the *Ignorabimus-Streit. Journal of the History of Philosophy, 56*(3), 519–549.

Street, S. (2006). A Darwinian dilemma for realist theories of value. *Philosophical Studies, 127*(1), 109–166.

Wimsatt, W. C. (2001). Generative entrenchment and the developmental systems approach to evo-
    lutionary processes. In S. Oyama, R. D. Gray, & P. E. Griffiths (Eds.), *Cycles of contingency:
    Developmental systems and evolution* (pp. 219–237). MIT Press.
Wimsatt, W. C., & Griesemer, J. R. (2007). Reproducing entrenchments to scaffold culture: The
    central role of development in cultural evolution. In R. Sansome & R. N. Brandon (Eds.),
    Integrating evolution and development: From theory to practice (pp. 228–323). MIT Press.

Chapter 6
Odum and Nietzsche: Parallels, Differences and Implications

Abstract This chapter examines the relation between the work of Friedrich Nietzsche and H. T. Odum: it discusses the many parallels I find in their work, the differences, and it considers the implications of this investigation. The parallels between the ontological version of the will to power and the maximum power principle are discussed in some detail, along with their use of these principles to critically analyze moral and religious values.

6.1 Introduction

Who would have thought that a nineteenth century German philosopher who was trained as a philologist and struggled throughout his life for recognition would have so much in common with a twentieth century American systems ecologist who had an enormously successful professional career? They certainly had different views on many important issues, which we will discuss, but they also had a surprising level of agreement on many fundamental issues, metaethical issues, such as the perspective we should use to critically evaluate moral and religious values.

This chapter explores the many parallels between the work Friedrich Nietzsche and H. T. Odum: it looks at their systems approach to their work; their use of science to better understand nature, morality, and philosophy; their view of nature as continually changing; and their view of all natural systems developing through evolutionary processes guided by the ontological version of the will to power (OWP) or the maximum power principle (MPP). This chapter will discuss how the OWP and the MPP apply to both biotic and abiotic systems, entail the three senses of transformation discussed in Chap. 5 (Sect. 5.2.1), affect our view of free will, describe efficiency being sacrificed to maximize power in certain situations and not in others, apply to human social systems as well as other natural systems, describe the growth of power as increasing human freedom, and how both operate in a manner that is structured by the pulsing paradigm.

This chapter discusses how Nietzsche critically analyzes moral and religious values from the perspective of the OWP, and Odum critically analyzes them from the

© The Author(s), under exclusive license to Springer Nature
Switzerland AG 2025
T. McWhirter, *Maximum Power and its Philosophical Roots*, SpringerBriefs in
Energy, https://doi.org/10.1007/978-3-031-80622-3_6

perspective of the MPP. It explains how Nietzsche and Odum both: describe good and evil as important to the functioning of social system; argue that in order to maximize power, patterns of behavior must "fit the earth;" describe some religions undermining the growth of power and some fostering it; and believe cultures provide important opportunities to sublimate drives and tendencies that could undermine the growth and flourishing of life.

Along with these parallels, there are many differences between the work of Nietzsche and Odum; in this chapter, we consider a few important examples: they had different views on the value of war and democracy and the nature of the evolution of moral and religious values. These differences can help us better understand the nature of the perspective they share in their critique of moral and religious values, and they can enable us to see inconsistencies or limitations in their respective positions.

Finally, this chapter will consider some of the implications that can be drawn from this investigation that relate to contemporary science, Nietzsche scholarship, and philosophy more generally. It will review the relation that has been described in earlier chapters between the MPP and the maximum entropy production principle (MEPP), and between the MPP and the minimum entropy production principle, and the evidence we now have that supports the MPP. It will describe the implications this investigation has for our understanding of Nietzsche's science of morals, his relation to general systems theory, his metaethical epistemology, his view of the slave rebellion in morality, and the empirical plausibility of the OWP. This chapter will end with a discussion of the implications this investigation has for philosophy more generally, beyond Nietzsche scholarship, and for life after the peak of our ability to use fossil fuels.

6.2 Parallels

6.2.1 General Systems Theory

One of the most basic and profound parallels between the work of Nietzsche and Odum is their systems approach to their work. This is a topic that is worthy of more consideration than I will be able to provide here, but this brief summary should be able to provide a basic understanding of this parallel and its significance.

In modern science, there has been a focus on understanding systems by analyzing their parts in detail. For example, there has been an effort to isolate the elements of the periodic table and understand the particles that make up atoms. This process is referred to as *analysis*. In contrast, general systems theory, an approach that evolved during the 1950s, focuses on synthesis rather than analysis and seeks to understand how the parts of systems interact and evolve within the context of their environment. Instead of analyzing the parts of systems in an isolated fashion, there is an effort to put the parts together and better understand how they work collectively as a system. This approach is referred to as *synthesis*.

In both ancient and modern philosophy, philosophers have used an approach to answer philosophical questions called *conceptual analysis*, which focuses on investigating the logical implications of philosophical concepts. Not all philosophers use this approach, but many have. This approach is often used to distinguish what is unique about philosophy and the work that philosophers do. In the twentieth century, there came to be a distinction between the kinds of philosophy that were practiced. *Analytic* philosophy was generally practiced by English and American philosophers. It generally focused on the analysis of logic, language, thought, knowledge, and the mind, among other things—it was focused on *analysis*. *Continental* philosophy, on the other hand, was generally practiced by German and French philosophers and it was focused on the synthesis of individuals with society and the present with the past—it was focused on *synthesis*. Many have argued that this analytic/continental dichotomy is problematic in many ways: it is, for example, simplistic and it does not characterize the work of many philosophers accurately. These concerns can be accurate and well taken, even though this dichotomy does accurately characterize the work of many philosophers in the nineteenth, twentieth, and twenty-first centuries.

Nietzsche is a German philosopher that took a decidedly *synthetic* approach to his work. In fact, he was trained as a philologist, someone who studies the history of languages. In order to better understand how languages evolve, a philologist often reads all the different kinds of literature written during different time periods, including the philosophy, science, as well as the novels and poetry. In his philological studies, Nietzsche focused on ancient Greece, as reflected in his first book *The Birth of Tragedy (BT)*. Nietzsche took this same philological approach to philosophy: he did not just read the work of philosophers; he read the work of biologists, physicists, historians, novelists, poets, and politicians, among many others. He synthesized all that he read in order to develop his philosophy. Thus, his philosophy is uniquely broad, but also unsystematic. It was written using aphorisms of various lengths that would often hop from one subject to the next. This, I believe, limited his ability to successfully argue for many of his unique views; however, what his approach did provide was a unique view of philosophy that synthesized the work from different disciplines—he could provide a view of the big picture. Ludwig von Bertalanffy found aspects of system theory in the work of the philosophers Gottfried Leibniz and Nicholas of Cusa. Others trace systems theory back further to the pre-Socratic philosopher Heraclitus (Hammond, 2003). Nietzsche had the highest respect for Heraclitus (*TI* Reason 2). It appears that Nietzsche was also, in his way, one of the precursors to modern general systems theory. We can see this in his view of human beings, the interdisciplinary nature of his work, his view of systems as the product of the interaction of subsystems, and his focus on power.

Many ancient and modern philosophers viewed human beings as being unique in some qualitative manner that is philosophically significant and separated them from other animals and forms of life. Aristotle thought that (male) human beings had a rational part of the soul that distinguished them from other animals. Nietzsche disagreed; he thought human beings were not different from other animals or forms of life in any philosophically significant manner, and he attempted to persuade his readers to give up any idea that human beings have some special goal or purpose that is different from other forms of life—in his view, all natural systems seek to grow (BGE 230). Nietzsche's view of the relation between human life and other forms of nature is therefore generally consistent with systems ecology. Traditional approaches to ecology had excluded the study of human social systems and focused just on ecological systems. In systems ecology, scientists seek to understand (among other things) the relation between human social systems and ecological systems. This idea informs Odum's investigation of human social systems in *Power, Environment and Society* (1971).

In his own work, Nietzsche was influenced by the work of philosophers throughout the history of philosophy, poets, artists, and a number of scientists: the English naturalist, geologist, and biologist, Charles Darwin, the Swiss botanist Carl Nageli, the German biologist Wilhelm Roux, the German Entomologist William Henry Rolph, the German physician and scientist of thermodynamics Julius Robert Mayer, and the Croatian physicist, astronomer, mathematician, philosopher, diplomat, poet, theologian, and Jesuit Priest, Roger Joseph Boscovich, among others. Nietzsche's philosophy is uniquely interdisciplinary.

General systems theory and systems ecology are interdisciplinary. Systems ecology includes physics, biology, ecology, and economics. In his famous 1955 paper on the MPP, Odum worked with the physicist Richard Pinkerton. Together they were able to apply the MPP to a number of different mechanical and electric systems, as well as physiological, ecological and economic systems. Odum's research over his long career included work that related to cosmology (2007, p. 269), anthropology (2007, ch. 8), economics (2007, ch. 9), psychology (2007, p. 299), morality (2007, p. 326; 1977), and philosophy. Like Nietzsche, his work was uniquely interdisciplinary.

In general systems theory, all systems are viewed as being composed of interacting subsystems. Each subsystem is itself a system composed of other subsystems. And all systems are viewed as evolving in relation to their environment. Mark Brown said that Odum would usually encourage his students to consider the behavior of systems with the context of the scale of phenomena that is larger than the system, and the scale that is smaller than the system: How does a forest relate to the atmospheric patterns of the earth? How does it relate to the molecular structure of the leaves? In his argument for the OWP, Nietzsche uses Boscovich's view of matter, which holds that matter is made up of points without dimension that exert fields of force. Nietzsche suggests this view provides a more accurate view of the soul: instead of viewing the soul as a monad, he thought it should be viewed as "a society constructed out of drives and affects" (*BGE* 12)—a system composed of subsystems. Nietzsche then goes on to argue that since this approach appears to explain matter and the soul, we should consider whether it works for all natural systems, both biotic and abiotic. Nietzsche's argument for the OWP (*BGE* 36) essentially concludes that we should take a general systems approach to understanding all natural systems.

Contemporary systems ecology investigates the flow of energy, nutrients, material, and information though natural systems. It focuses on understanding systems that are far from equilibrium, in the sense that they are continually receiving energy from their environment. These systems are dissipative structures: their organization is sustained by dissipating energy from their environment. Nietzsche sought to understand the flow of power and information through social systems and the relation between the two. This flow is the focus of the OWP. The third sense of transformation entailed in the OWP, discussed in Chap. 5 (Sect. 5.2.1), illustrates that Nietzsche, on some level, was aware of the dissipative structure of natural systems: He described all centers of force growing through the "appropriation and assimilation" of external forces (*WP* 656).[1]

6.2.2   Nietzsche, Odum and Naturalism

Both Nietzsche and Odum analyze moral and religious values critically and say they do so from the perspective of science. Nietzsche wrote that we should seek to understand human beings the same way we seek to understand other parts of nature, from a perspective "hardened by the discipline of science" (*BGE* 230). The previous chapter described how he did this.

However, Nietzsche's commitment to science is not without qualifications. He thought there were limits on how it can be used effectively. He writes:

> Assuming that one estimated the value of a piece of music according to how much of it could be counted, calculated, and expressed in formulas: how absurd would such a "scientific" estimation of music be! What would one have comprehended, understood, grasped of it? Nothing, really nothing of what is "music" in it! (*GS* 373)

Efforts to understand everything from a scientific perspective have been referred to by philosophers as *scientism*; Nietzsche was clearly opposed to this view. He believed that we should be "hardened by the discipline of science," but also respect the limits of its domain. His discussion of the "science of morals" clearly demonstrated that he thought morality was within the domain of science (*BGE* 186).

Nietzsche also thought there were limits on what we can learn from science when it is applied to phenomena within its domain. He outlines a distinction between explanation [*Erklärung*] and description [*Beschreibung*]: "'Explanation' is what we call it, but it is 'description' that distinguishes us from older stages of knowledge and science. Our descriptions are better—we do not explain any more than our predecessors" (*GS* 112; Soll, 2018, p. 532). He believes science can provide valuable descriptions of natural phenomena which can continually be improved, but it cannot provide a true explanation for why phenomena occur. In this sense, his views are consistent with those of Julius Robert von Mayer. Mayer made it clear that developing equations that describe the transformation of motion into electricity accurately does not explain the physical process by which these transformations take place. He suggests that questions regarding how to explain the physical processes behind

---

[1] Nachlass 1885–1887, 9[151], *KSA* 12.424.

these transformations "are useless and typical of poets and philosophers of nature" (Mayer, 1978, p. 10; Coelho, 2009, pp. 965–966).

Nietzsche's view is not consistent with the distinction Lotka makes between descriptive and exact science in his essay "The Law of Evolution as a Maximal Principle" (1945, p. 172). Lotka suggests that "the exact sciences look beyond mere description. They aim at establishing coherent disciplines within which, by the application of relatively few fundamental principles, the course of events can be rigorously deduced for innumerable specific situations." He believes the fundamental laws of thermodynamics are examples of such fundamental principles. Nietzsche would argue that these "exact sciences" may provide more accurate descriptions of natural phenomena, but they do not provide true explanations for why the phenomena occur.

Nietzsche writes that the "typology of morals" that is needed for a "science of morals" should be based on "the task of description" (*BGE* 186), and this task can reveal the "real problems of morality; for these emerge only when we compare many moralities." He also compares moralities as they develop over time in a genealogy. This approach enables him to uncover signs of the will to power at work (*GM* II 12). Nietzsche's naturalism is therefore based on a descriptive view of scientific practice.

For his part, Odum often found himself in the position of pointing out how people often overlook the extent to which science can help them better understand the phenomena they see and their own behavior. The wars that occur, the behavior of economies, the moral and religious values people have can all be better understood if we recognize their relation to the flows of energy through social systems and the fundamental thermodynamic laws. We have no reason to think Odum believed that science can help us better understand the beauty of music, but he did continually press the boundaries others have placed on the domain of science. In his critique of morality and his development of a "science of morals," Nietzsche was doing the same thing.

6.2.3 Change Is Essential to Nature

As mentioned, Nietzsche had the highest respect for Heraclitus, including how he described change and multiplicity as being essential to nature at a time when philosophers were arguing that nature is based on permanence and unity (*TI* "Reason" 2). Heraclitus had to argue for this on his own, something with which Nietzsche was certainly familiar. Nietzsche suggests that since nature is always changing, we can improve our understanding of it if we seek to describe in detail how it changes over time, through a genealogy (or taxonomy). Those that have this kind of knowledge— those who can see the "farthest back and ahead" (*A* 57)—are in a better position, in his view, to make a decision about what we should do now.

In his analysis of natural systems, Odum often focused on describing energy transformations, i.e., changes in the energy cycling through systems. He described patterns in how things change over time, such as successional or evolutionary time. He understood that nature is continually changing and in order to better understand it, we need to improve our understanding of how it changes over time. Although the energy systems language he developed is of necessity static, it is in reality a snapshot describing the movement of energy over time: it is a language of change.

Both Nietzsche and Odum accepted Darwin's view that life evolves over time through a process of selection. They viewed this evolutionary process as guiding the development of both biotic and abiotic systems. They both also believed that this evolutionary process was guided by something more fundamental than survival and reproduction.

6.2.4 The MPP and the OWP

Evolution

This section will attempt to discuss the parallels between the MPP and the OWP, an ambitious task. The first parallel between them that we will consider is that they are both described as guiding the evolution of natural systems. Darwin described evolution being based on the mechanism of natural selection, which provides a selective advantage to organisms that have heritable traits that enable them to survive and reproduce more effectively than their competitors. Ludwig Boltzmann viewed evolution from a thermodynamic perspective and suggested that the 'struggle for existence' was really a struggle for low entropy. Alfred Lotka further developed this approach, arguing that those organisms (sometimes Lotka says systems) that are able to direct more energy through the processes associated with survival and reproduction will have a selective advantage. Odum and Pinkerton develop Lotka's insight further and describe in more detail how organisms and natural systems that can direct more useful power output through the processes associated with survival and reproduction will have a selective advantage. In this way, the process of natural selection was redefined from the perspective of thermodynamics as being guided by the MPP.

For Nietzsche, this change in the understanding of evolution is not characterized as being merely a redefinition from the perspective of thermodynamics; he is critical of Darwin's description of the evolution of life being guided by a "struggle to exist" (*TI* "Skirmishes" 14) or an "instinct of self-preservation" (*BGE* 13). He often misinterprets "self-preservation" as a tendency to remain the same, to not change, which is not what Darwin meant. For their lineage to be preserved, Darwin knew that organisms must evolve and develop traits that enable them to improve their ability to survive and reproduce, especially as the environment changes. So, part of Nietzsche's argument against "self-preservation" is based on a strawman fallacy. Nietzsche believed in evolution, but he believed that it was guided by a struggle for more power, not a struggle to merely exist. He did not appear to see that if we view the struggle to exist from a thermodynamic perspective, the two options begin to get closer to each other: those organisms that are able to direct more useful power output through the processes associated with survival and reproduction will have a selective advantage. In any case, Nietzsche describes evolution being guided by the OWP, and Odum describes it being guided by the MPP.

Abiotic and Biotic Systems

The OWP and the MPP are both described as applying to abiotic systems as well as biotic systems. In his argument for the OWP, Nietzsche describes it applying to "a more primitive form of the world of affects in which everything still lies contained in a powerful unity before it undergoes ramifications and developments in the organic process …—as a pre-form of life" (*BGE* 36). In his notes, Nietzsche writes: "what has been the relation of the total organic process to the rest of nature? That is where its fundamental will stands revealed" (*WP* 691).[2] He sees the OWP as this fundamental will: it is what the "rest of nature" shares with the "total organic process." Odum cites the work on streams by Sugita (1951) and Leopold and Langbein (1962) as providing some empirical support for the MPP (Odum & Richardson, 1982, p. 118, 1983a, p. 118), and he also argues that the MPP applies to hurricanes and concentrated masses (stars) in space (2007, p. 90).

Three Forms of Transformation

In Chap. 5, I discussed how Nietzsche described the OWP as being transformative in three different ways: systems transform themselves; they transform systems in their environment; and they transform systems in their environment in a way that generates the entropy necessary to sustain the development of their own organization over time as required by the second law of thermodynamics. The MPP also entails these three senses of transformation. The energy systems language Odum developed and used illustrates how these three senses of transformation are involved in the development of all natural systems. It illustrates how energy passes through systems, how energy is taken from the environment, and the heat sinks highlight the third sense of transformation: energy from the environment is dissipated, increasing entropy, enabling natural systems to develop their organization in a manner that is consistent with the second law.

"Free Will"

The OWP and the MPP both have an impact on the concept of "Free Will." Odum would often say that people have the freedom to make their own choices, but only those choices that are consistent with the MPP will have a selective advantage in evolutionary processes; consequently, in the future, there will be more people that make choices consistent with the MPP. In this way, systems can draw from individuals the behavior that maximizes power (Odum, 1977, p. 111). It happens through a process of cultural evolution and most people are unaware of the fact that we, and our "free choices," are all products of it. Our perception that we are exercising our "Free Will" is itself part of the larger evolutionary process, generating the choices from which natural selection chooses.

---

[2] Nachlass 1885–1886, 2[99], *KSA* 12.109. References to Nietzsche's unpublished works included in the *KSA* will include the term Nachlass, the year of the note, the fragment number, *KSA*, the volume number, a decimal point, and then the page numbers.

Nietzsche's critique of "free will" is well known in the secondary literature. We covered part of it in his argument for the OWP. He rejects the "atomism of the soul" that characterizes it as an individual monad unilaterally responsible for making choices, and he argues that this view should be replaced with "new versions and sophistications of the soul hypothesis," such as the "soul as a society constructed out of drives and affects" (*BGE* 12). This new version describes decisions as the products of interactions between different drives and Nietzsche thought individuals are not consciously aware of many of these interactions. After one drive comes to dominate the others, the individual then comes to be aware that a decision has been made and, after the fact, claims responsibility for it. It is like a rooster that crows at dawn and then claims to be responsible for the sun rising. For Nietzsche, the idea of "free will" was a creation that enabled societies to punish people for their behavior (*GM* II 7; *EH* Destiny 8). In the interactions between the drives in the society of the soul, Nietzsche saw the will to power at work. The OWP and the MPP both lead us to qualify the view that our choices are the products of an unfettered "free will."

Efficiency and Power
Nietzsche and Odum both describe a unique relation between efficiency and power. Odum's view has been discussed at some length in Chap. 4. In the ideal case where there is no limit to the available energy, the optimal efficiency for maximum power will never exceed 50% (1955, p. 333; fig. 2). In this ideal case, efficiency is sacrificed to maximize power. However, as the available energy and resources are reduced in cases that are not ideal, the optimal efficiency for maximizing useful power output is increased and it can go as high as 90% (1955, p. 341, 343; 2007, p. 214). In these cases, maximizing efficiency will maximize power.

Nietzsche's view of the relation between efficiency and power, to the extent to which he has one, makes more sense if we keep Odum's position in mind. In the early 1880s, Nietzsche's position appears to be based on his reading of the entomologist William Henry Rolph and the physician and thermodynamic scientist Julius Robert Mayer. Rolph described evolution being guided by a struggle for the *expansion* of life, which wasted more energy than would have been used in order to merely preserve life. Mayer investigated the energy wasted in all kinds of energy transformations: by cannons, steam engines, and various animals. Nietzsche saw Mayer's work as providing empirical support for Rolph's view that the struggle for the expansion of life can come at the expense of efficiency (Emden, 2014, p. 173). In his notes from 1881, Nietzsche quotes Mayer in an approving manner: "The chemical process is always larger than its useful effect."[3] Nietzsche's rejection of Darwin's view that evolution is based on a struggle to preserve life is, in part, based on the idea that the struggle is for more power, and this, in many cases, comes at the expense of efficiency.

On the other hand, in 1887, Nietzsche also writes the following in his notes: "That which constitutes growth in life is an ever more thrifty and more far-seeing economy, which achieves more and more with less and less force—As an ideal, the

---

[3] Nachlass 1881, 11[24], *KSA* 9.451; Mayer, 1945, p. 102 & 116–18.

principle of the smallest expenditure" (*WP* 639).[4] We should remember that we are focusing here on two passages that Nietzsche wrote in his notes; this is not material that he published. Six years passed in between these two entries. In these two passages, he does not mention anything about the energy available in the environment. So, as it stands, these two passages could be interpreted as contradicting each other. It would not be the first contradiction philosophers have found in Nietzsche's ideas (Müller-Lauter, 1999).

However, Odum's description of the relation between efficiency and the MPP suggests that a more accurate interpretation of these two passages is possible. They could be interpreted as signs Nietzsche was in the process of working out this relation between efficiency and power. He appeared to be aware that sometimes power could be maximized by sacrificing efficiency, and that in other situations, power could be maximized by increasing efficiency. According to Odum, Nietzsche is correct on both counts and Odum explains further how this can be the case. So, we could conclude that Nietzsche was on the right track; he was just unable to completely figure out the relation between efficiency and OWP.

Human Power

As mentioned in Chap. 5, the MPP and OWP both use the concept of power to refer to the power manifested by human beings. As Odum acknowledges, this is unique; our education system and scientific disciplines do not generally use the scientific concept of power to apply to the power manifested by human beings, even though many different courses talk about political and military, power. The fact that the MPP and the OWP both apply to the power manifested by human beings is a unique and important parallel between them.

Nietzsche and Odum both describe the growth of power increasing the amount of freedom that human beings are able to enjoy. In Chap. 5, we reviewed a passage from *Beyond Good and Evil* where Nietzsche referred to the growth in power of the process of production and how it enabled people to "become increasingly independent of any determinate milieu that would like to inscribe itself for centuries in body and soul with the same demands" (*BGE* 242). Odum describes societies with "high-energy flows" as providing "more choices," which cause the patterns of behavior to become more "fluid" (1977, p. 131). When these energy flows are reduced, the choices for individuals are reduced, and the patterns of behavior are less fluid. Some philosophers and political scientists have associated the growth of power with the growth of repression and the limitation of freedom. There certainly are individual instances where this is the case, just because societies have the ability to provide individuals more freedom does not mean they will always do so. However, if we stand back and consider the choices humans had before the Neolithic revolution and compare them with the choices we have now, there is no comparison: We can see a dramatic increase in freedom, an expansion of what Emden describes as the "scaffolding" of power.

---

[4] Nachlass 1887, 10[138], KSA 12.535.

The Pulsing Paradigm

The OWP and the MPP are both described as being structured by a pulsing paradigm. For the MPP, this was described in Chap. 4, Sect. 4.6. The development of natural systems is structured by a pulsing rhythm on multiple temporal and spatial scales: the seasons and the days; at the molecular and planetary level. Odum argues that this pulsing rhythm can maximize power (Odum & Richardson, 1981). For Nietzsche, this pulsing rhythm is evident in his description of the tempo of the reevaluation of values that fosters the growth and flourishing of life as outlined by the OWP.

Nietzsche believed that in order for societies to grow and flourish, they must have the "strength to will, and to will something for a long time…." (*BGE* 208). He refers to this ability as the "tension of the spirit" (*BGE* Preface). In order to manifest this tension, a society must be able to manifest widespread agreement on (or acceptance of) goals and be able to persist until the goals are achieved. Nietzsche is critical of two tempos in the reevaluation of values that undermine this tension: one is synchronic; the other is diachronic.

First, we will consider the synchronic component. Nietzsche describes how the struggle with difficult obstacles can lead a society to create individuals that are strong and hard; a society without such challenges can lose the tension that creates strength. It can lead to the creation of a,

> kind of *tropical tempo* in the rivalry of growth, and an extraordinary decay and self destruction, owing to the savagely opposing and seemingly exploding egoisms, which strive with one another 'for sun and light', and can no longer assign any limit, restraint, or forbearance for themselves by means of the hitherto existing morality. (BGE 262, my emphasis)

Nietzsche provides an implicit stipulative definition for the term "tropical" here that is probably at odds with the understanding of ecologists who study the intricate organization of ecological systems, like Odum. Nietzsche uses the term "tropical" to refer to a lack of collective organization: where plants are fighting amongst each other "for sun and light" rather than working together toward a collective goal. The "savagely opposing" interests in this *tropical tempo* undermine the ability of a society to gain widespread agreement on (or acceptance of) important goals.

This "tension of the spirit" is also weakened by a "*prestissimo* tempo" in the revaluation of values. This concern represents the diachronic component of the structure of the "tension of the spirit." Nietzsche describes "Modernity" in his notes using the metaphors of nourishment and digestion:

> Sensibility immensely more irritable (—dressed up moralistically: the increase in pity—); the abundance of disparate impressions greater than ever: cosmopolitanism in foods, literatures, newspapers, forms, tastes, even landscapes. The tempo of this influx prestissimo; the impressions erase each other; one instinctively resists taking in anything, taking anything deeply, to 'digest' anything; a weakening of the power to digest results from this. A kind of adaptation to this flood of impressions takes place: men unlearn spontaneous action, they merely react to stimuli from outside." (Nachlass, 1887, 10[18], *KSA* 12.458; *WP* 71)

Prestissimo is the fastest tempo in music. When people regularly change their minds quickly, it undermines a society's ability to persist until its long-term goals are achieved. In order for a society to have the "strength to will, and to will something for a long time….," it must be able to avoid the *tropical* and *prestissimo* tempos in the revaluation of value (*BGE* 208).

The tempo of the reevaluation of values that Nietzsche implicitly calls for is *largo* (very slow) and *symphonic* (in the form of an organized symphony), not *prestissimo* and *tropical*. The members of a society must be able to work together over an extended period of time to achieve long-term goals. Once these goals are achieved, then societies must reevaluate their values and set new goals. This tempo of reevaluation is consistent with the pulsing paradigm. For example, since the middle of the nineteenth century, societies have been using fossil fuels to achieve unprecedented levels of growth. However, Odum recognized that as the peak of our ability to use fossil fuels approaches, societies will eventually have to reevaluate their values and adopt practices that are better adapted to a situation in which there is less available energy (Odum & Odum, 2001). Odum sees this reevaluation as a part of the pulsing paradigm. It will enable societies to maximize the power that is available under these new conditions. Changes like this are not made by societies every other Tuesday; they are made after long periods in which societies have struggled to achieve goals. As mentioned, we see this pulsing rhythm in the concept of punctuated equilibria (Eldredge & Gould, 1972, 1993). We also see it in the *largo* and *symphonic* tempo of Nietzsche's "tension of the spirit," which is a fundamental part of the will to power manifested by human social systems.

6.2.4.1 Morality and the OWP and the MPP

Not only do Nietzsche and Odum critically analyze moral and religious values from the perspective of science, but they both do so using these two principles that have all these similarities: the OWP and the MPP. They consider whether the values under consideration maximize power more than the alternatives. Odum provides an "emergy test for morality" and an "energy system ethics for all scales" (2007, p. 327, 329).

At the outset of Nietzsche's *On the Genealogy of* Morals (*GM* Preface 3), he makes the focus of his critique of morality clear: "under what conditions did man devise these value judgments good and evil? and what value do they themselves possess? Have they hitherto hindered or furthered human prosperity? Are they a sign of distress, of impoverishment, of the degeneration of life? Or is there revealed in them, on the contrary, the plenitude, force, and will of life, its courage, certainty, future?" He already made it clear in *Beyond Good and Evil* that he understood the growth and flourishing of life as being outlined by the OWP (*BGE* 13, 259), and, as illustrated in Chap. 6, he saw himself as describing the affects moral and religious values have on this growth (*BGE* 186).

The Necessity of Good and Evil
Nietzsche and Odum both describe good *and* evil being necessary in order to maximize power. Nietzsche writes that,

> We think that hardness, forcefulness, slavery, danger in the alley and the heart, life in hiding, stoicism, the art of experiment and devilry of every kind, that everything evil, terrible, tyrannical in man, everything in him that is kin to beasts of prey and serpents, serves the enhancement of the species "man" as much as its opposite does (*BGE* 44).

That evil which does not kill us, makes us stronger, more resourceful and resilient, better able to handle whatever comes our way (*TI* Maxims 8). He writes that "the great epochs of our life come when we gain the courage to rechristen our evil as what is best in us" (*BGE* 116). When we set aside the common assumptions about the importance of altruism and we seek to accomplish our own goals for our own reasons, we can further the progress of humanity.

Nietzsche's view here is supported by his rejection of moral dualism or what he calls "*the faith in opposite values*," like good and evil (BGE 2). He adopts a scientific monism, based on the OWP (*BGE* 2; Richardson, 2015). From the perspective of this monism, any set of values can maximize power more than some other set of values and less than some others. Therefore, judgments about moral values are always relative to the alternatives. A set of values may have maximized power more than the alternatives in Nietzsche's time, but in Odum's time, the situation may have changed: people could be aware of alternatives that they were not aware of in Nietzsche's time. So, that which is good today, may have been evil in the distant past or ignored and vice-versa. For example, the Christian moral values Nietzsche criticized in his own time must have served at some point in the past to maximize power more than the alternatives, otherwise they would not have existed in the nineteenth century.

Odum writes that order is often associated with good and the highest moral values; disorder is often associated with the devil and evil (1977, p. 114). In his systems approach, he points out that both order and disorder are fundamental parts of all natural systems: he describes them being in a symbiotic relationship (1977, p. 113, Fig. 4). An example is the decay of dead organic matter in a forest that makes nutrients again available for new growth. This relationship is necessitated by the second law of thermodynamics: in order for the organization of a natural system to increase over time, the entropy of its environment must increase. From this perspective, we can see evil as a "flow into disorder" that is necessary to sustain the dissipative processes that generate the organization of natural systems. Therefore, both order and disorder, and good and evil, are necessary from a systems perspective.

Fit the Earth
In their analysis of morality, both Nietzsche and Odum write that the patterns of human behavior need to "fit the earth." In *Thus Spake Zarathustra*, Zarathustra tells us to "remain faithful to the earth…" (Z Prologue 3), and not pay attention to those who preach of otherworldly hopes, for example, the hopes of an eternal life in heaven. The message here is that fostering the growth and flourishing of life on the earth is the only real option we have; the otherworldly hopes are based on illusions that distract us from this task.

Odum describes how maximizing power requires fitting cultural patterns to the earth and the conditions in the environment (2007, p. 329, 392). One of the values in his energy system ethics is "fit the earth." Human societies in the twenty-first century are reaching a point where changes need to be made in order to adapt to conditions where there is less available energy. We need to become more efficient, self-sufficient, and reduce population growth. This will enable our culture to "fit the earth" as it is now and make it possible for us to maximize power to the extent that we can under the present conditions.

Religion and Growth

Nietzsche and Odum both describe some religions as fostering the growth and flourishing of life and the maximization of power, and they describe others as undermining this growth. Nietzsche spends the majority of his time criticizing religions for undermining the growth of life. He is well known for his blistering critique of Christianity, which is the focus of his book *The Anti-Christ*. But, in this book, you will also find him describing the Laws of Manu, the Hindu law book, as providing an outlook that fosters the growth of life (*A* 57). These examples are generally hard to find. Throughout his work, Nietzsche tends to focus on his critique of different views, rather than those areas where he is in agreement with them. This is particularly the case with respect to his analysis of religion.

Odum provides a unique energy analysis of religion (1971, ch. 8; 1977; 2007, ch. 11). He describes the different ways that religions can play a crucial role that enable societies to foster the growth and flourishing of life by maximizing their power; he also describes how religions can undermine this growth by maintaining approaches that do not fit the earth and the existing conditions. He discusses different religious pathologies where societies have too little or too much religion (2007, p. 317). He actually uses his energy systems language to diagram the "essence of the soul" in the "chain of energy quality" (1977, p. 122). He describes a rather idealistic vision of how religious leaders can become more aware of issues relating to energy and the environment and they can use this knowledge to adapt their religious teachings. One thing that you find in Odum's analysis of religion that you do not generally find in Nietzsche's analysis is an awareness that religions evolve over time (2007, p. 323). New types of societies emerge along with new types of religions. This appears to be one of the reasons that Odum's analysis of religion is more optimistic.

Sublimation

Nietzsche and Odum also describe how contemporary societies provide different ways that individuals can sublimate different drives that could otherwise undermine the growth and flourishing of life. Nietzsche writes the history of higher culture has been marked by an "ever increasing spiritualization and 'deification' of cruelty" (*GM* II 6). "That 'savage animal' has not really been 'mortified;' it lives and flourishes, it has merely become—divine" (*BGE* 229).

> Perhaps the possibility may even be allowed that this joy in cruelty does not really have to have died out: if pain hurts more today, it simply requires a certain sublimation and subtilization, that is to say it has to appear translated into the imaginative and psychical and adorned with such innocent names that even the tenderest and most hypocritical conscience is not suspicious of them ('tragic pity' is one such name; 'les nostalgies de la croix' [the nostalgia of the cross] is another). (*GM* II 7)

Nietzsche describes higher culture translating this joy into many different forms.

> What the Roman in the arena, the Christian in the ecstasies of the cross, the Spaniard at an auto-da-fe or bullfight, the Japanese of today when he flocks to tragedies, the laborer in a Parisian suburb who feels a nostalgia for bloody revolutions, the Wagnerienne who 'submits to' Tristan and Isolde, her will suspended—what all of them enjoy and seek to drink in with mysterious ardor are the spicy potions of the great Circe, 'cruelty'. (*BGE* 229)

All drink the spicy potions, but do so in different ways. Today, people have a taste for violent sports, movies, and video games. Cultural processes make it possible to sublimate this joy in cruelty in many very different ways, from the arena, stadium, theater, to the living room.

Odum describes sporting events providing a relatively harmless outlet for some of the aggressive and competitive behavior that could undermine the functioning of contemporary societies. These sporting events are like "storms" that can absorb these aggressive and competitive tendencies that might otherwise be manifested in destructive forms of violence and war (2007, p. 299). In 1907, Allan Sangree of the *New York World* described baseball in a similar and prescient manner that can apply to all sports:

> The fundamental reason for the popularity of the game is the fact that it is a national safety valve. Voltaire says that there are no real pleasures without real needs. Now a young, ambitious and growing nation needs to "let off steam." Baseball furnishes the opportunity. Therefore, it is a real pleasure…. That is what baseball does for humanity. It serves the same purpose as a revolution in Central America or a thunderstorm on a hot day…. A tonic, an exercise, a safety-valve, baseball is second only to Death as a leveler.

If baseball is second only to Death as a leveler, what is contemporary American football? We see a discussion of this kind of sublimation in Sigmund Freud (1930), who was himself influenced by Nietzsche, and similar processes have also been described by contemporary behavioral ecologists (Brown & Laland, 2006).

Most importantly, both of these authors saw that natural selection had left subliminal aggressive and acquisitive traits in humanity that caused humans to strive toward greater control of their environment and the resources therein. Without explicitly endorsing the virtue or vice of this trait itself, each believed it important to acknowledge and understand that as one key to understanding ourselves and our species.

6.3 Differences

Nietzsche and Odum had different views about a number of things. These differences make the parallels between them, stand out in contrast. They had very different approaches to the way they communicated their ideas, dealt with opposing views, and lived their lives. In this section, I will be focusing on the differences in the positions they took on war, democracy, the nature of the evolution of morality, and the status of their critique of morality. These differences can help us better understand the nature and significance of the issues upon which they agree.

6.3.1 War

Nietzsche, who in his youth spent some time in the military, had a view of war that was consistently and alarmingly sanguine. He criticizes every philosophy that "ranks peace above war" (*GS* 2) and he writes that "what is generally overlooked is that the ancient national energy and national passion that became gloriously visible in war and warlike games …" (*GS* 23). Nietzsche sees war as a "remedy" for cultural exhaustion: it forces societies to build and gather together an unprecedented amount of energy and power in order to survive.

> It is vain reverie and beautiful-soulism to expect much more (let alone only then to expect much) of mankind when it has unlearned how to wage war. (…) Culture can in no way do without passions, vices and acts of wickedness.—When the Romans of the imperial era had grown a little tired of war they tried to gain new energy through animal baiting, gladiatorial combats and the persecution of Christians. Present-day Englishmen, who seem also on the whole to have renounced war, seize on a different means of again engendering their fading energies: those perilous journeys of discovery, navigations, mountain climbings, undertaken for scientific ends as they claim, in truth so as to bring home with them superfluous energy acquired through adventures and perils of all kinds. One will be able to discover many other such surrogates for war, but they will perhaps increasingly reveal that so highly cultivated and for that reason necessarily feeble humanity as that of the present-day European requires not merely war but the greatest and most terrible wars—thus a temporary relapse into barbarism—if the means to culture are not to deprive them of their culture and of their existence itself.[5]

Nietzsche's view of war is therefore related to his view of the OWP: he sees war as a way for cultures to grow and become more powerful.

To understand what Nietzsche is talking about, consider that WWII led countries to develop a new form of science: Big Science—where scientists from different disciplines are brought together and their activities are collectively organized in order to accomplish a particular goal. This new approach to science was used in WWII to create the first primitive computer that was used to break the German code for information, the first radar systems, and the first atomic weapons. The war effort also in many countries provided a decisive end to the great depression.

According to Nietzsche, we are indebted to Napoleon because he led us into a few warlike centuries that have no parallel—what Nietzsche describes as the *the classical age of war*:

> *Our faith that Europe will become more virile*—We owe it to Napoleon (and not by any means to the French Revolution, which aimed at the "brotherhood" of nations and a blooming universal exchange of hearts) that we now confront a succession of a few warlike centuries that have no parallel in history; in short, that we have entered *the classical age of war*, of scientific and at the same time popular war on the largest scale (in weapons, talents, and discipline). All coming centuries will look back on it with envy and awe for its perfection. (*GS* 362)

---

[5] The translation is provided by Benjamin Biebuyck (2018, p. 170). *KSA* 2.311–312, aphorism 477.

Nietzsche's view of war remained largely consistent throughout his life. In *Twilight of the Idols*, originally published in 1888, he writes, "One has renounced the great life when one renounces war" (*TI* Morality 3).

Lotka, writing during World War II, was taken aback by the carnage and death caused by the war (1945, p. 189). The fact that people were killing each other in another world war and birth rates were going down in many countries below the level necessary to sustain a population was interpreted by Lotka as evidence that contemporary human civilizations were no longer developing in a manner that is guided by his *principle of maximum energy flux*. He did however note that WWII led to the discovery of atomic energy and he suggested that if this discovery were put to constructive use, it would be a "superlative example" of how the principle of maximum energy flux guides evolutionary development. So, his view of WWII was mixed.

Odum's view of war was also mixed. He served as a lieutenant in WWII. But, when we consider the difference between his view and Nietzsche's within the context of the time Odum was writing, one could argue that his position is closer to Nietzsche's than we might initially think. Odum suggests that in the early stages of a society's development, wars can be useful for the purpose of "territorial organization and renewal" (2007, p. 304). In times of low energy, wars can help expand societies and clarify a division of resources (2007, p. 307). In these times, war can channel energy into organization.

However, Odum described the situation being different in the second half of the twentieth century. In this high energy period, the power of the weapons that can be used in war had increased to a point where they can do unprecedented damage to social systems and their environments; the extraordinary quantities of energy used in military programs and campaigns were not always translated into increases in organization; they were often simply wasted or used in ways that undermine the growth of power. Odum writes that in his time, international agencies had emerged to control conflict and avoid the damage that can be caused by war. He suggested that emergy evaluations are needed to determine whether this international approach maximizes power more than the decentralized power politics that involve periodic wars (2007, p. 307).

When we attempt to understand the difference between Nietzsche's and Odum's positions, there is one thing that stands out: the first atomic weapon was used in WWII. In a sense, this fact both illustrates Nietzsche's point, and demonstrates why it no longer applies. First, the invention of atomic weapons illustrates how war can enhance the development of power. But, second, it also illustrates that in this atomic (and now nuclear) era, war can not only undermine the development of power, it can bring about the mutual destruction of all involved in war. During the cold war, the nuclear strategy for the United States and the Soviet Union was literally called MAD: mutually assured destruction. Each side had enough nuclear weapons to destroy the other country. As a result, neither side had an incentive to launch a first strike because it would be suicidal.

In this nuclear age, Nietzsche's logic with respect to war no longer applies, if indeed it ever did. He argued that war can be a remedy for cultural exhaustion: it can increase a society's power. In the nineteenth century, that might have been the case, but in the nuclear age, that logic can lead a society to its annihilation. The adaptation to Nietzsche's view that is called for in this new environment is already explicitly stated in his work. He describes an "ever-increasing spiritualization and 'deification' of cruelty" taking place, an ever increasing sublimation of violent drives (*GM* II 6). Rather than going to war on the battlefield, we can, in an ever-increasing fashion, go to war in philosophy, religion, sports, the law, science, music, film, and business. Nietzsche did this himself in an ever-increasing fashion: late in his life, he began to use the rhetoric of war more and more in his writings. In the *Anti-Christ*, he calls for a war against Christianity, and the traditional concepts of "true" and "untrue" (A 13). These "wars" can continue in this nuclear age and they can foster the growth and flourishing of life as outlined by the OWP. So, if we take into consideration this important difference—Odum is writing in a nuclear age—the different views Odum and Nietzsche had on the value of war are easier to understand.

6.3.2 Democracy

Nietzsche and Odum provide different descriptions of the value of democracy. Nietzsche describes it as undermining a society's ability to grow and flourish as outlined by the OWP. In *On the Genealogy of Morals*, he discusses how the "prevalent instinct and taste" is opposed to acknowledging that "in all events a *will to power* is operating," and he then writes,

> The democratic idiosyncrasy which opposes everything that dominates and wants to dominate, the modern misarchism (to coin an ugly word for an ugly thing) has permeated the realm of the spirit and disguised itself in the most spiritual forms to such a degree that today it has forced its way, has acquired the right to force its way into the strictest, apparently most objective sciences; indeed, it seems to me to have already taken charge of all physiology and theory of life—to the detriment of life, as goes without saying, since it has robbed it of a fundamental concept, that of activity. (*GM* II 12)

We get a better understanding of his concern here if we consider comments he makes earlier in *Beyond Good and Evil,* where Nietzsche uses a metaphor of a bow and arrow to refer to the "tension of the spirit" that I discussed previously in Sect. 6.2.4. Those societies that are able to manifest widespread agreement on fundamental goals and persist until these goals are fulfilled are able to exert more power than those that cannot do this. These societies are described by Nietzsche as having a great "tension of the spirit," and he compares it with the tension of a bow that enables societies to "shoot for the most distant goals" (*BGE* Preface).

Nietzsche writes that in Europe, there have already been two attempts to "unbend the bow"—

> once by means of Jesuitism, the second time by means of the democratic enlightenment which, with the aid of freedom of the press and newspaper-reading, might indeed bring it about that the spirit would no longer experience itself so easily as a "need." (The Germans have invented gunpowder—all due respect for that!—but then they made up for that: they invented the press.) (*BGE* Preface)

Here, democracy is described by Nietzsche as undermining the "tension of the spirit:" reducing a society's ability to manifest widespread agreement on long-term goals and maintain this agreement for a long period of time. He associates democracy with the prestissimo tempo in the revaluation of values: as quoted earlier, "cosmopolitanism in foods, literatures, newspapers, forms, tastes, even landscapes. The tempo of this influx prestissimo; the impressions erase each other; one instinctively resists taking in anything, taking anything deeply, to 'digest' anything" (*WP* 71).[6] He appears to associate newspapers with democracy: they provide information to citizens that influences their votes. So, for Nietzsche, democracy undermines the tension of the spirit and thereby undermines the growth and power of social systems.[7]

Odum does an energy analysis of democratic and totalitarian political systems. The main difference he describes between them is that in democratic systems, people have the ability to vote on their leaders and this allows them to reinforce the controls that work (2007, p. 295). In totalitarian systems, he suggests that there is no mechanism for keeping the leaders effective or efficient in supporting the system and its production. As a result, totalitarian systems are not "self-correcting," and this could cause revolutions to periodically occur. By comparison, democratic systems are self-correcting, and this reduces the likelihood of revolutions. Odum believed that the governments that are able to prevail over time control the most power (2007, p. 295).

We will consider the significance of the differences in Nietzsche's and Odum's analysis of democracy later in this chapter when we discuss the implications of this investigation.

6.3.3 The Evolution of Morality

Both Nietzsche and Odum describe the evolution of life being guided by the OWP or the MPP. Odum acknowledges throughout his work that this fact applies to the evolution of moral values. He describes how the moral values in the United States during the energy-rich late twentieth century did not place the same importance on restricting sexual promiscuity as was the case in primitive societies that had limited access to energy (2007, p. 327). He argues that this explains why the attempts to remove president Clinton in 1999 were unsuccessful: the moral standards of the time had evolved along with the increase in the availability of energy. When Odum discusses the consequences of the eventual decline in our ability to use fossil fuels, he describes there eventually being an evolution in our moral values to correspond to this change in conditions (2001).

While Nietzsche clearly describes life evolving in a manner that is guided by the OWP (*BGE* 13, 259), he did not describe the evolution of morality in the same

---

[6] Nachlass 1887, 10[18], *KSA* 12.458.
[7] For an opposing view, see Hatab (1995).

manner; in fact, he often describes the evolution of morality as heading in a direction that undermines the OWP. Perhaps the clearest illustration of this is his description of the *slave revolt in morality*. He writes that "with the Jews there begins *the slave revolt in morality*: that revolt which has a history of two thousand years behind it and which we no longer see because it—has been victorious" (*GM* I 7). The "slave morality" that emerges from this revolt stands in opposition to the "master morality" that was prevalent in ancient Greece (*BGE* 260). Nietzsche characterizes the *slave morality* and the *master morality* as being two types of moralities. The *slave morality* originated from those who have been ruled by others, the slaves: it seeks to make life easier for slaves; it holds that all are created equal and it ensures that everyone is provided equal consideration and an equal ability to be happy. The *master morality*, on the other hand, originated from those who have ruled over others, the masters: it does not view people as being equal; it holds that there is an order of rank that justifies the privilege of the rulers; and it acknowledges that strict moral values are needed for societies to grow and flourish. Nietzsche argues that the slave morality has come to dominate in his time and, as a result, the ability of societies to grow and become more powerful has been undermined.

Nietzsche's claims about the slave morality contradict his view that the evolution of life is guided by the OWP, and they are not supported by the empirical evidence reviewed in Chap. 5. Over the course of the 18th and 19th centuries, societies increased their ability to manifest power in a nonlinear fashion. The first industrial revolution occurred between 1760 and 1840; the second between 1870 and 1914. Nietzsche wrote almost all his work during the second industrial revolution. He acknowledged how the growth in the power of the productive process in societies in his time increased the freedom people could enjoy (*BGE* 242).

Nietzsche's discussion of the slave and master moralities provides us resources that can be used to overcome this contradiction in his position. Nietzsche acknowledges that in "higher and more mixed cultures," there "appear attempts at mediation between these two moralities," and they can even mix together within one person (*BGE* 260). While Nietzsche was a young man, the United States had a civil war in order to mediate between versions of these two types of moralities. We see a form of this mediation in the political parties in societies around the world: liberals or progressives that support equality and equal rights lean toward a slave morality; conservatives that seek to reduce taxation and regulation on the wealthy leaders of industry and fight efforts to support equality lean toward a master morality. Nietzsche himself seems, at times, to acknowledge this mediation. In the preface to one of his last books, *The Anti-Christ*, he acknowledges that the audience for his work is quite small: "This book belongs to the very few. Perhaps not one of them is even living yet. Maybe they will be the readers who understand my Zarathustra…" He outlines a number of conditions which must be met in order for people to understand his work and writes: "Such men alone are my readers, my right readers, my predestined readers: what matter the rest? The rest—that is merely mankind. One must be above mankind in strength, in loftiness of soul—in contempt." He appears to realize that he is writing to the rulers, the *ubermenschen*, the leaders that follow the master morality, and the rest of mankind is not paying attention.

Some form of mediation between the master and slave moralities could enable societies to increase their power more than societies that use the master morality by itself. Imagine, for example, a society where most people abide by a slave morality of one form or another; the leaders of the society give lip service to it, but in their mind, they are guided by a master morality. In his notes, Nietzsche describes the ruled going through a process of "dwarfing and adaptation" that is necessary to develop a "specialized utility" that serves needs of the economy (*WP* 866). They spend their lives working with this "specialized utility." The slave morality utilized by most people could make the life of the ruled much easier to bear: it could give them the impression that all people are really equal; some people are just luckier than others—they are born into wealthier families, or they have had good fortune in business. The rulers, on the other hand, what Nietzsche calls the "synthetic, summarizing, justifying" men, use the economy to invent their "*higher form of being.*" In a society where there is a mediation between the slave and master morality, the rulers could act and think in ways that are guided by the master morality and they could publicly show their support for the slave morality to the extent they need to in order to do what they want and to get others to do what they want. This public support for a slave morality can serve to increase the loyalty of the ruled and their willingness to work hard.

Nietzsche appears to assume that societies can manifest more power if there is a universal acceptance of a master morality. His view is not supported by the empirical evidence and it is inconsistent with his own description of evolution being guided by the OWP. We can explain the fact that the power of social systems has grown since the "*slave revolt in morality*" by acknowledging that a "mediation" between the master and slave moralities appears to have enhanced the ability of societies to increase their power (see, e.g., Curtis, 2022).

Nietzsche's concern about the threat of nihilism provides another example of how he did not fully understand the implications of his own view that evolution is guided by the OWP. Near the end of his academic career, Nietzsche planned to complete a comprehensive philosophical project; the first book in this project was entitled, "Nihilism, Completely Thought Through" (Nachlass, 1888, 12[2], *KSA* 13.211). He was concerned that in the wake of the decline in religion, which he saw coming, the threat of nihilism would emerge as a fundamental challenge to the growth and flourishing of life.

By comparison, Odum also saw a threat to the growth and flourishing of life coming in the future—the peak of our ability to use fossil fuels. He was so concerned about this that he wrote a book about it: *A Prosperous Way Down* (2001). However, in this book and others, he described how our moral and religious values would evolve in a manner that helps us deal with this challenge. We generally do not see this same recognition in Nietzsche's work. He was clearly aware that there was a process of cultural evolution at work (*GM* II 17), but he did not appear to have the same confidence in it as Odum.

One lesson to be learned here is that the processes involved in cultural evolution function collectively in ways that individuals in these cultures do not fully understand. In this sense, we are all like the members of the tribes in Fiji that do not fully understand the significance of their food taboos. In order for us to critically analyze moral values in a manner that actually fosters the growth and flourishing of life, as outlined by the MPP and OWP, we must seek to better understand why the moral values under consideration exist in the first place; the fact that they do demonstrates that they have served to foster the growth and flourishing of life at some point. Second, we need to not assume that we fully understand the effects of these moral values. This kind of humility might not be Nietzsche's strong suit.

6.3.4   The Metaethics of the Science of Morals

In Chap. 5, Sect. 5.3, I described how Nietzsche views the OWP as an empirical principle rather than a moral value, and he uses it as a foundation for a thermodynamic framework for the development of social systems. He describes the effects that moral and religious values have on the growth and flourishing of life from the perspective provided by this framework. He makes it clear that his descriptions of these effects are not moral claims: he does not argue that certain moral values are immoral; he essentially describes moral values as either undermining or fostering the growth and flourishing of life. Because he does not attempt to define the OWP as a fundamental moral value, his position is not vulnerable to the "open question" argument; it is supported by it.

Odum believed that since the laws of thermodynamics help explain the development of social systems, these laws "define what is moral" (2007, p. 327). This claim could raise some questions for philosophers: is Odum suggesting that these laws define what is moral in a necessary manner given the meaning of the words "moral" and "laws of thermodynamics"? Is he suggesting that the question—Do these laws actually define what is moral?—is closed? I believe the most accurate way to interpret Odum's claim is as an empirical hypothesis, which remains open, like all other empirical hypotheses. He gathered evidence that demonstrates moral and religious values evolve in a manner that can be described using the laws of thermodynamics. Based on this evidence, he is suggesting that what is moral appears to be defined by the laws of thermodynamics.

Philosophers will point out that in ancient Greece, women were not treated as being equal to men, and this empirical fact cannot be used to conclude that it is moral to treat women this way. This point is well taken. However, from the perspective of a science of morals, if we have evidence that throughout human history moral values have evolved in a manner that can be explained using the laws of thermodynamics, then we have good reason to think this will continue to be the case in the future. The question will always be open, strictly speaking, but the evidence gives us reason to have confidence in our description, just like we have confidence in our description of gravity.

In Nietzsche's vernacular, Odum is not arguing that the MPP is another "rational foundation for morality;" he is outlining a science of morality that is based on an empirical foundation: the MPP. He was always open to the possibility that future developments in science would provide a better understanding of the things he discussed in his work. Odum's effort to re-define "what is moral" and Nietzsche's effort to reconsider the value of morality (GM Preface 6) are both made from the perspective of science, which mitigates, to a considerable degree, the significance of the difference that remains between their positions.

This difference in the approaches Odum and Nietzsche take that remains is however interesting: Odum attempts to redefine what is moral in a naturalistic manner, while Nietzsche re-evaluates the value of morality itself from a naturalistic perspective. Ironically, Odum, the scientist, appears to be working within the language game of morality, while Nietzsche, the philosopher, uses a scientific perspective to question our faith in morality itself. They both describe moral and religious values playing an instrumental role that fosters the growth and flourishing of life; consequently, we are not in a position to claim that either approach is true or more accurate. It is really a matter of what is more effective at influencing people to behave in ways that enable social systems to grow and flourish, as outlined by the OWP and the MPP. In fact, it might be more effective to use both approaches; one could be used with some groups and another could be used with other groups: Odum's approach could be more effective with philosophers; Nietzsche's might be more effective with scientists.

6.4 Implications

Now that we see the many parallels and the differences between the work of Nietzsche and Odum, what are the implications of this investigation? What conclusions can we draw? What kind of impact can this investigation have on contemporary science, Nietzsche scholarship, and the practice of philosophy? What kind of lessons can it provide us about life after the peak of our ability to use fossil fuels? This last section attempts to answer these questions.

6.4.1 The MPP and the MEPP

The maximum power principle (MPP) and the maximum entropy production principle (MEPP) are fundamentally related. Any time a natural system increases its power, it also increases its ability to produce entropy. So far, there has not been a detailed analysis that explains the relation between these principles and considers their relative value. I believe this would be an important and valuable project.

The fact that the majority of the contemporary scientists doing work on the development of complex systems tend to use the MEPP to describe the development of systems far from equilibrium gives us reason to think this principle is particularly valuable. However, as Charles Hall has noted, Lotka and Odum argue that systems that direct more *useful* power output through the processes associated with survival and reproduction have a selective advantage in evolutionary processes. The production of entropy is also selective for, but only incidentally because of its fundamental relation to useful power output. Consequently, it appears that the evolution of natural systems is actually driven by the MPP. So far, no one has provided any answer to this basic theoretical point.

6.4.2   MPP and the Minimum Entropy Production Principle

Sciubba describes contemporary scientists using two principles to describe the development of natural systems: systems near equilibrium tend to evolve in a manner that maximizes efficiency, as described, for example, by the minimum entropy production principle; systems far from equilibrium tend to evolve in a manner that maximizes the dissipation of energy, as described by the MPP or the MEPP.

However, Sciubba (2011) fails to acknowledge the fact that Lotka and Odum wrote that the optimum efficiency needed to maximize useful power output increases as the available energy decreases. Odum argues that this will maximize the power that is "available" under these conditions (1982, p. 35). Consequently, the suggestion that systems near equilibrium develop in a manner that maximizes efficiency is not an exception to the MPP; it is consistent with it (1982, p. 35). The MPP outlined by Odum and Lotka's PMEF are therefore able to accurately describe the development of all natural systems, both near and far from equilibrium.

6.4.3   The Empirical Evidence for the MPP

The arguments that Lotka offered for the principle of maximum energy flux were based on theoretical considerations; he did not provide empirical evidence to support the principle. Odum cited a couple of empirical studies that supported the MPP, but they were clearly not the focus of his argument for the MPP. He admitted that it is extremely difficult to test the principle directly.

Over the past few decades, the situation has changed dramatically. In the studies done by Lenton et al. (2016), Steffen et al. (2015), and Thimsen (2024), we find historical evidence that the energy cycling through natural systems increases over time in a manner that is consistent with the MPP and MEPP. This evidence provides a challenge to some traditional interpretations of evolutionary theory that characterize it as a process that is, for the most part, random. The philosopher Thomas Nagel's argument that this traditional interpretation of evolution is unable to provide a compelling explanation for the existence of brains and consciousness illustrates that we all have access to additional evidence that supports the MPP and the MEPP: our brains and our experience of consciousness (2012). These brains made it possible for swordfish and whales to devise complex cooperative processes to trap prey, as well as human beings to continue to develop machines that were more and more powerful, propelling the growth of the exosomatic use and dissipation of energy that led at the extreme to the development of nuclear power.

Sciubba suggested in 2011 that there is no proof that demonstrates the PMEF or the MPP accurately characterizes the development of natural systems (2011, p. 1351). He was unable to discuss any of the evidence I just mentioned: the situation has, indeed, changed.

6.4.4 The Evolution of the "Science of Morals"

Nietzsche and Odum critically analyze moral and religious values from a perspective that is surprisingly similar: they both analyze these values from the perspective of science, and, more specifically, from the perspective of the OWP and MPP. Notwithstanding this fact, they come to have different views about the value of war and democracy, just to name two of the most important differences. What are we to make of these differences? What do they illustrate about the "science of morals" implicit in Nietzsche's and Odum's critical analysis of moral and religious values?

First of all, both Nietzsche and Odum analyze moral and religious values from the perspective of science, rather than some philosophical concept of moral truth that is ahistorical. The doctors that treat diseases that undermine the growth of life in the twenty-first century use approaches that are very different from those used by doctors in the twentieth and nineteenth centuries. The science of medicine, like all sciences, is continually developing: studies and experiments are being done, new therapies are being developed and used, more and more data is being accumulated. Consequently, doctors today are far better able to treat illnesses than doctors could a century ago. Nietzsche and Odum are engaged in a critique of moral and religious values that has fundamental similarities to the practice of medicine: they describe the effects these values have on the growth and flourishing of life and they prescribe therapies that foster this growth. Nietzsche's description of himself as a cultural physician is more than merely a metaphor. Given the scientific nature of their approach, we should expect to see differences in Nietzsche's and Odum's analyses, and we should also expect to see ways their analyses can be improved using the knowledge and information we have now or may have in the future. For example, our perception of the value of democracy will change over time as we are able to gather more information about it and how it enables social systems to address the challenges they face relative to the available alternatives. We have learned a great deal about democracies, totalitarian governments, and other political systems since the nineteenth century, and this experience can contribute to our understanding of their value. We learn a great deal about this during our own lives. Odum's students observed his willingness to change his perspective over time, from support and then criticism of the Vietnam war. This fact that we learn more about things over time is a fundamental part of the evolving science of morals outlined in the work of Nietzsche and Odum.

Second, with respect to their views on war, there was a development in WWII that clearly had a transformative impact on the nature of war—the creation of atomic weapons. Nietzsche saw war as a way that cultures could remedy exhaustion and stimulate growth; Odum saw the same potential when considering primitive societies (2007, p. 304, 307). But, when we consider mature societies in a nuclear age, the potential damage that can be caused by an unlimited war fundamentally changes the logic of Nietzsche's position: such a war could cause the destruction of all the societies involved and their environments—all growth could come to an end. These new situations, of which Nietzsche was not aware, force us to reinterpret the implications of his position. In his philosophy, he describes a continuing process of sublimation that leads contemporary human beings to sublimate their aggressive and violent drives through processes that do not undermine the growth and flourishing of life—bullfights, tragedies, the opera etc. In this nuclear age, the growth of the power of social systems is dependent on their ability to sublimate the drives that could lead them to an unlimited nuclear war. We must recognize this possibility in order to accurately interpret the implications of Nietzsche's philosophy in our nuclear era.

Third, the value monism discussed earlier illustrates that the moral and religious values that maximize power now will eventually undermine the growth of power in the future because alternatives will emerge that can maximize power more. There is no limit to the growth of power; therefore, there is no limit to the evolution of the values that foster the growth of power. This aspect of the evolution of morality can also explain why Nietzsche, Odum, and we now may have different views about the effects moral values have on the growth and flourishing of life.

This investigation therefore helps us better understand the differences between Nietzsche's and Odum's critical analyses of moral and religious values and how they are (or at least appear) consistent with the scientific perspective they share. This investigation also illustrates how Nietzsche's argument for his "science of morals" can be dramatically improved.

Chapter 5 explains how Nietzsche uses his argument for the will to power (discussed in Sect. 5.2.2) and his problem of morality itself argument (Sect. 5.3.3) to support his version of a "science of morals" that is based on an empirical foundation: the OWP (*BGE* 186). Now, we have the ability to use Lotka's and Odum's theoretical argument for the MPP to support the case for a "science of morals" based on an empirical principle: the organisms and natural systems that are able to direct more useful power output to the processes associated with survival and reproduction will have a selective advantage in evolutionary processes. This theoretical argument for the MPP is far more compelling than Nietzsche's argument for the OWP. It, among other things, explicitly provides the mechanism by which it works: natural selection. We can also include the empirical evidence for the MPP discussed in Chap. 4 (Sect. 4.10). We also have the ability to use G. E. Moore's "open question" argument to further undermine the assumption that there is a "rational foundation for morality" (discussed in Sect. 5.3.5). And we can include Sharon Street's argument that evolution has had a profound influence on our evaluative attitudes and realist interpretations of moral value are unable to provide a plausible explanation for this influence (2006). This argument illustrates that evolutionary theory provides a more compelling explanation for our evaluative attitudes than realist interpretations. Together, these arguments and this empirical evidence make a compelling case for a science of morals based on an empirical foundation: the MPP.

Arguments that Support a Science of Morals
1. Lotka's and Odum's theoretical argument for the MPP
2. Nietzsche's problem of morality itself argument
3. Moore's "open question" argument
4. Street's evolutionary influence argument

Nietzsche's version of this "science of morals" will have a different focus than traditional approaches to moral theory. It will focus on providing *descriptions*, rather than answering normative questions: it will focus on describing how things are rather than prescribing how they ought to be. It will *describe* the effects moral and religious values have on the growth and flourishing of life as outlined by the OWP. It will *describe* how these values evolve over time in a manner that is guided by the OWP. Based on these descriptions, this "science of morals" will be able to *describe* the nature and function of moral and religious values. Odum's approach will appear to address normative issues more directly, but it will do so from the perspective of a "science of morals" based on a foundation that is empirical rather than rational: the MPP. Since the thermodynamic laws that "define what is moral" are descriptive, one can argue that Odum science of morals is also fundamentally descriptive (2007, p. 327).

The idea of a "science of morals" can bring to mind legitimate concerns about a kind of "technocratic optimism" that need to be taken seriously (see, e.g., Taylor, 1988). Going forward, this science of morals needs to be developed through a critical dialogue that seeks to understand the issues and problems that have been overlooked at various stages, like the democratic systems Odum describes, which, he argues, are more stable than a "simple dictatorship" (1971, p. 212–213). In keeping with general systems theory, an interdisciplinary approach will be needed.

That being said, this concern about "technocratic optimism" needs to be joined with a concern about the *"analytic optimism"* that has dominated discussions of morality and moral theory in philosophy since Plato: the idea that we can best understand the meaning and function of moral values through the process of conceptual analysis, e.g., by examining the logical implications of the definition of the term "goodness." Some philosophers, like Thomas Nagel, believe the importance of these conceptual issues cannot be undermined by anything we may learn from psychology, biology, or the study of cultural evolution, because the "normative cannot be transcended by the descriptive" (1979, p. 106). This *analytic optimism* can overlook the importance of many issues and problems considered in the sciences, like global climate change. A "science of morals" that is informed by general systems theory can provide an important counterweight to this *analytic optimism*. It should be particularly valuable as we struggle to better understand the long term effects of human activities and their impact on global climate change.

We can see an example of a kind of "science of morals" in the contemporary sustainability movement. Right now universities, colleges, corporations, governments, and families all around the world are reducing their carbon emissions, among other things, in order to deal with the threat posed by global climate change. People of all ages, from senior citizens to children in elementary school, believe this is important. Like Nietzsche's and Odum's "science of morals," the contemporary sustainability movement views human behavior from a scientific perspective; it attempts to steer people away from forms of behavior that undermine the long-term growth of life; and it views the significance and meaning of human behavior and moral values in terms of their relation to the growth of life.

6.4.5   Nietzsche and General Systems Theory

I have not been able to find any studies of the relation between Nietzsche's philosophy and general systems theory. This is surprising because there are so many books on Nietzsche that focus on different aspects of his philosophy. General systems theory brings into relief a fundamental focus that guides Nietzsche's approach to philosophy: he is focused on synthesis; he views human beings within the context of a larger system; his approach to philosophy is uniquely interdisciplinary; he sees all systems as being composed of interacting subsystems; and he focuses on the energy and power manifested by systems. These connections between Nietzsche's philosophy and general systems theory are provocative and I have only had a chance to scratch the surface of them.

6.4.6   Nietzsche's Naturalistic Metaethical Epistemology

The chapter on Nietzsche develops a unique interpretation of his metaethical epistemology that helps us better understand his critique of morality, one of his most famous contributions to the history of philosophy. Reading lists in graduate studies programs in philosophy will generally include Nietzsche's book, *On the Genealogy of Morals*, even in departments that have a focus that is more analytic. The argument provided in this investigation is that Nietzsche does not criticize moral values from the perspective of other moral values; he criticizes them from the perspective of science: he *describes* the effects moral and religious values have on the growth and flourishing of life, as outlined by the OWP, understood as an empirical principle. His critique is, therefore, not based on moral claims or judgments; it is based on empirical descriptions. The metaethical epistemology implicit in his critique of morality is therefore best understood as a form of *empiricism*. His critical analysis of the "science of morals" practiced in his time provides a unique view of his perception of the "science of morals" he develops (*BGE* 186).

General systems theory provides a framework that can help us better understand the nature of Nietzsche's critique of morality. His critique does not come from within morality through some form of conceptual analysis; it comes from outside it, from science: biology, sociology, and thermodynamics. He sees morality as one part of a larger system, which is evolving over time as outlined by the will to power. He describes the effects moral values have on this development. This systems perspective provides a unique view of what morality is and how it functions.

6.4.7 Reevaluating the Slave Rebellion in Morality

Nietzsche's critique of the slave morality is a central element of his critique of morality, along with his critique of Christianity. This investigation demonstrates that Nietzsche's view of the "slave rebellion in morality" is not consistent with his view that the evolution of life is guided by the OWP and it is not supported by the empirical evidence: the first and second industrial revolutions demonstrate that the power of social systems increased during the 18th and 19th centuries in a nonlinear fashion, when Nietzsche described the "slave rebellion" continuing to develop.

We can interpret the significance of the "slave rebellion in morality" in a manner that is consistent with Nietzsche's view of evolution and the empirical evidence that is now available if we recognize the potential of societies that seek a mediation between the slave and master moralities. The evidence suggests that these societies have, in fact, fostered the growth and flourishing of life, as outlined by the OWP. This reinterpretation of Nietzsche's view enables us to remove a contradiction at the heart of his critique of morality.

6.4.8 The Empirical Plausibility of the OWP

The assumption that the OWP is empirically implausible according to the contemporary sciences has had an extraordinary impact on the secondary literature on Nietzsche. It has led influential philosophers to change the course of their research on Nietzsche; some have developed "esoteric" interpretations of his philosophy that suggest that he really did not mean what he said about the OWP; many have been led to adopt psychological interpretations of the will to power, which they believe are better supported by the contemporary sciences. None of the philosophers who have criticized Nietzsche's OWP for being empirically implausible have mentioned the MPP, the PMEF, the MEPP, or the work of Lotka or Odum or any of the scientists in the "thermodynamic school" of evolution theory, or any of the empirical evidence that we now have that supports the MPP and the OWP.

Odum did not begin to use the name—the *maximum power principle*—until 1976, and at that point, some scientists began to use the maximum entropy production principle instead of the MPP. The relation between the MEPP and the OWP would not be obvious to philosophers. This clearly appears to have contributed to the problem. Whatever the case may be, we are now in a unique position to recognize the exceptional prescience of an extraordinary philosopher. During his life, his ideas were ignored by the philosophical community. After his death, many of his ideas became more popular and were interpreted in different ways for different purposes, but his OWP has been consistently rejected and ignored by philosophers. One of the aims of this investigation is to demonstrate that we are now in position to give Nietzsche the credit he has long deserved for his OWP.

6.4.9   The Evolution of Knowledge

In these last two subsections of this chapter, we will briefly consider some implications of this investigation that relate to philosophy, beyond the study of Nietzsche. A comprehensive analysis of these implications would take us well beyond the scope of this particular book. The suggestions outlined here are intended to illustrate the different lines of research that could be pursued in order to better understand the implications of this investigation.

Odum describes moral values evolving over time in a manner that is guided by the MPP, but he does not explicitly describe scientific knowledge evolving in the same way. Like moral values, the scientific knowledge developed by social systems has a profound influence on their development; it is a cultural product, developed and shared through social interactions, used by members of that community, and it is therefore influenced by the processes of cultural evolution. If we follow the implications of Odum's work, we need to view scientific knowledge as evolving in a manner that is guided by the MPP.

The feedback loop described by Lotka that was discussed in Chap. 4, Sect. 4.5, provides a general picture of how scientific knowledge can evolve in a way that is guided by the MPP (1945, p. 188). Lotka describes human beings developing improvements in the process of production that provided them more time and energy that was available for "play" and luxuries, and some of this excess time and energy was devoted to scientific research. This had a dramatic impact on the development of social systems that resulted in the agricultural and industrial revolutions (1945, p. 188). These revolutions provided more human beings with more excess time and energy and a portion of it was devoted to scientific research. As mentioned, the philosopher of science Philip Kitcher argued that taking different approaches to scientific problems increases the probability that a solution will be discovered (1990). When more people are engaged in scientific work, it enables more of them to take different approaches to different scientific problems. We can see the effects of this process reflected in the graph of scientific discoveries and inventions provided by Lotka (Chap. 3, Sect. 3.4.1) and the historical table of heat engine technology provided by Earl Cook (Chap. 4, Sect. 4.10). Since the first industrial revolution, many believe that we now have had at least two more, the second between 1870 and 1914, and third between 1947 and 2009. These revolutions have enabled us to devote even more time and energy to scientific research.

Lotka illustrates how scientific knowledge has enabled human social systems to maximize their power in a way that has enhanced their ability to generate more scientific knowledge. Societies that maximize their power in this way more than others will be better able to share their scientific knowledge with other societies through the processes associated with intergroup competition in cultural evolution: war and raiding, differential group survival without conflict, differential migration, prestige-biased group transmission, along with other processes, such as skill-bias, success-bias, and conformist transmission (Henrich, 2016, pp. 167–168, 39, 47). Through these processes, the knowledge that enables societies to maximize power has a selective advantage in cultural evolution.

Nietzsche did not evaluate knowledge from the perspective of some metaphysical concept of truth; he evaluated it from a pragmatic perspective: is it "life-promoting, life-preserving, species-preserving, perhaps even species-cultivating..." (*BGE* 4). By promoting the growth and flourishing of life, knowledge has the power to foster the growth of power. Sir Francis Bacon and his one-time secretary, Thomas Hobbes, both believed that knowledge is power (Bacon, 1861, p. 94–95; Hobbes, 1668, p. 69). Lotka provides a brief illustration of how knowledge has the power to generate more knowledge.

6.4.10 The Evolution of Philosophy

In 2011, the philosophers David Morrow and Chris Sula published an essay which developed a provocative argument that philosophy departments serve as engines of influence that "steer" the development of philosophy as a discipline (2011, p. 301). They argued that this influence was largely the product of two particular psychological principles that worked together: uniformity pressure and confirmation bias. Uniformity pressure holds that when people spend a lot of time working together and interacting with each other, they tend to find things upon which they agree. Confirmation bias holds that once people believe something is true or accurate, they tend to seek and notice evidence and information that supports their belief.

Morrow and Sula cite the philosopher Gerald Allen Cohen for an example of the influence they describe. Cohen notes that the students who studied with him at Harvard tended to reject the analytic/synthetic distinction and he did not think this was an accident. He believed that the fact that he and his colleagues spent a lot of time with the professors at Harvard who rejected the distinction clearly had an impact on them (2000, p. 18).

Rational argumentation clearly plays a fundamental role in the development of philosophy as a discipline according to Morrow and Sula; they make this clear. But, they argue that uniformity pressure and confirmation bias also play a substantial role and that we should investigate their influence further. We can see their analysis outlining the ways that philosophical beliefs, such as the rejection of the analytic/

synthetic distinction, are reproduced in the next generation, like cultural memes. If we want to develop a more comprehensive understanding of how philosophy as a discipline is influenced by the processes associated with cultural evolution, we would consider the different ways the practice of philosophy is influenced by the other processes involved in intergroup competition mentioned above. For example, those societies that are better able to maximize their power will be better able to survive and flourish, and better able to support more philosophy departments that educate more philosophy students. The practices of the philosophers at the most prestigious universities in these societies will influence others in a disproportionate manner. Philosophy students from around the world will seek to be admitted to these prestigious institutions in a disproportionate manner. For these reasons, the philosophical practices of the professors at the most prestigious institutions in the societies that have maximized their power more than others will have a selective advantage. These processes, along with many other factors, will influence the development of philosophy as a discipline.

This brief sketch provides, at best, an outline for research that can be done to further develop the promising line of inquiry initiated by Morrow and Sula and explore in more detail the implications this investigation has for the development of philosophy.

6.4.11   Life After the Peak

Nietzsche recognized how societies in his time had grown in a "tremendous" way and this enabled people to exercise an unprecedented level of freedom in how they lived and worked (*BGE* 242). As a result of his historical vantage point and his training as a scientist, Odum was able to see that the growth of industrial societies continued to increase in a nonlinear fashion after Nietzsche's death, and this growth was fueled by a particularly powerful and efficient source of energy: fossil fuels. This source of energy is finite. Odum recognized that when our ability to use it begins to diminish, it will have a profound impact on societies. Scientists now believe that the peak of our ability to access and utilize fossil fuels has approached or is approaching quickly. The charts below (Fig. 6.1) provides some of the latest projections (See also Laherrère et al., 2022). They illustrate that the past 174 years have been a unique period in human history: Industrial societies have grown in an unprecedented manner and the people in them have become accustomed to a way of living that is fueled by an exponentially increasing use of fossil fuels. We have used this energy to build our homes, businesses, schools, stadiums, roads and highways, cities, culture etc. What will life be like on the other side of the peak? How will this change influence the evolution of our moral and religious values?

We know from Odum's description of the MPP that the optimum efficiency for maximizing useful power output increases as the available energy decreases: so efficiency will have a selective advantage in evolutionary processes after the peak, efficiency in public policy, business strategies, and moral and religious values, among other things. Odum's description of the evolution of moral and religious values in systems that are declining will apply here: these values will no longer focus on consuming luxuries; they will begin to focus on education, location efficient lifestyles, birth control, efficient infrastructure, personal responsibility for health, and respect for diversity. When the quantity of available energy decreases and the energy return of investment (EROI) decreases, the cost of this energy will generally go up. Since this energy is used to produce and deliver essentially all products, the price of products will generally go up, creating inflation. This inflation, together with the recognition that the society is no longer growing, can create widespread dissatisfaction that leads to political turmoil, which can enhance the political division between those that seek security in the religious and moral values of the past and those that are eager to embrace new values that are better able to address the present and future challenges. In his book *A Prosperous Way Down*, Odum provides a detailed analysis of the challenges we face after the peak of our ability to access and utilize fossil fuels (2001) (Fig. 6.1).

Odum interpreted the peak of fossil fuel use as a part of his pulsing paradigm. He viewed contemporary civilization as an "unsustainable pulse" that consumes the "fuel storages of the earth faster than they are replaced" (2007, p. 124). He thought "the increased carbon dioxide from this consumption pulse threatens" ecological systems around the world. He described human history as a series of pulses "of progress." And he suggested that the "pulse of the fuel-civilization of our time" will recede after the peak into "a lower-energy state." We can see the pulsing rhythm in systems throughout the natural world. Odum described how the prey-predator oscillation described in the Lotka-Volterra model can serve in "ecosystems as a pacemaker (analogous to a heart pacemaker) to provide an appropriate internal pulse" (2007, p. 155; see the diagram below). In this model, we see a cycle of growth and decline. Odum viewed the peak of our use of fossil fuels as the point at which societies will transition from growth to decline.[8]

In *A Prosperous Way Down*, Odum also considered the views of many who predict that there will be continued expansion and growth in the future. Some of them suggest that this continued growth will be fueled by developments in nuclear energy technology (Cetron & O'Toole, 1982; Simon & Kahn, 1984; Celente, 1997), increases in efficiency (Naisbitt, 1985; Repetto, 1985), and an increasing use of renewable energy (Fuller, 1968). Odum does not share their optimism, and he is not alone.

Odum argued that plants using nuclear fission require large investments to operate safely and the emergy yield ratio (the emergy yielded divided by the emergy invested) is approximately 4.5, less than half that of Alaskan oil (11.1) or coal from Wyoming (10.5) (2001, Table 10.1). Nuclear power also presents a number of

---

[8] Also see Hall and Detwiler (1988). There are cycles of destruction and rebirth in nature: forest cycles, delta cycles etc. Joseph Tainter has argued that such cycles (or growth and collapse) of individual cultures occur on a regular basis as cultures or empires operate in a way similar to maximum power until they get increasingly difficult to maintain as supply lines, food requirements, military discipline and so forth become to complex and hence energy costly to maintain (1988).

S.H. Mohr et al / Fuel 141 (2015) 120–135

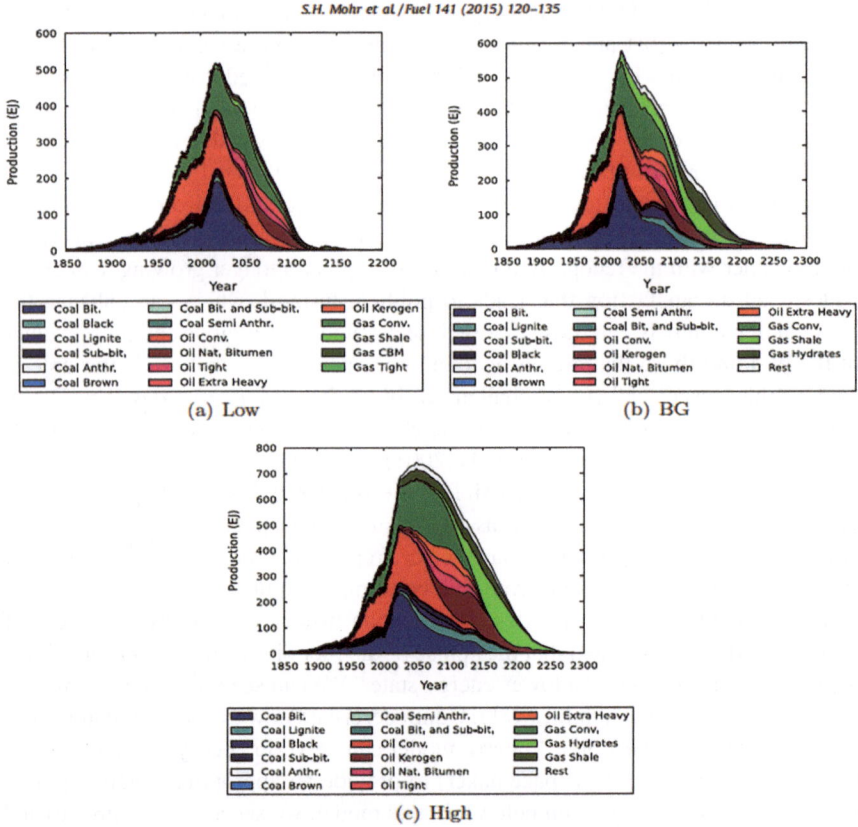

Fig. 6.1 Fossil Fuel projections by mineral type (Mohr et al., 2015)

unique risks and dangers. He pointed out that we are a long way from being able to generate energy on an ongoing basis using nuclear fusion. He argued that solar voltaic cell electricity requires an investment of more emergy than it yields; the emergy yield ratio is 0.4. While solar energy captured by green plants will always be a fundamental part of human life, the use of solar energy technology will not be an efficient or practical way to replace our present use of fossil fuels when all costs are considered. The emergy yield ratio of electricity generated from wind energy technology is approximately 2, far below that of fossil fuels, and it is only possible to generate this energy in certain locations at certain times. Odum believed that energy efficiency will become an even more important part of life after peak oil as we transition to a low-energy society, but increases in efficiency will not enable contemporary industrial societies to continue their present rate of growth without the use of fossil fuels.

That being said, the more positive predictions Odum mentions do appear to be supported to some degree by some of the findings in this investigation, like the scientific feedback loop described by Lotka. As a result of the three industrial revolutions we have already been through, we now have more scientists doing different kinds of research on energy and energy technology than we have had at any point in human history. As mentioned in Chap. 5, the philosopher of science Philip Kitcher argued that taking different approaches to scientific problems increases the probability that a solution will be found (1990). So, there is a higher probability that we will find a solution to our energy dilemma now than there ever has been in the past.

Also, the energy revolutions that have occurred in human history do not follow the pulsing pattern.

	Begins approximately
Use of fire	1.5 million BCE
Neolithic revolution	12,000–10,000 BCE
1st industrial revolution	1760 AD
2nd industrial revolution	1870 AD
3nd industrial revolution	1947 AD

The pattern of these revolutions is different from the pulsing pattern described by Odum in at least two respects. First, each revolution is not followed by a downturn that eventually settles into a steady state; they are followed by a nonlinear increase in the energy flowing through human social systems. This is confirmed by the work of Lenton et al. (2016) and Steffen et al. (2015). Second, these revolutions do not occur at regular intervals, like the days, tides and seasons; the time of these intervals is decreasing. The rate of change is increasing in speed suggesting a form of acceleration, as described by the historian Henry Adams.

We see the pulsing pattern throughout human life, in the oscillation of the work humans do each day and each season, the pulses of work done by humans over the course of their life, and the pulses of growth that occur in different societies at different times. But the energy revolutions that have occurred through human history do not fit the pulsing pattern; they fit a pattern of acceleration. So, if the peak of our ability to use fossil fuels does in fact lead to a downturn that eventually settles into a low-energy steady state, it will be the first energy revolution that has done this in human history.

The accelerating pattern of energy revolutions outlined above can, however, fit within a larger pulsing pattern that is consistent with Odum's descriptions. Scientists have come to recognize five major mass extinction events over the past 500 million years, which they call the "Big Five." A number of scientists have recently argued that there is a sixth mass extinction event that we will need to add to this list: we are living in this event now. They call it the *Holocene Extinction Event* (Briggs, 2017; Cowie et al., 2022; Bradshaw et al., 2022). Holocene is the name given to the last 11,700 years of earth history. This mass extinction event is primarily attributed to the activity of human beings. The rate of extinction of species right now is estimated to be 100 to 1000 times higher than the historically typical rate of extinction. The "Great Acceleration" described by Steffen and his colleagues provides evidence of the impact human activities are having on ecological systems around the world (2015).

The "Big Five" extinction events		
1.	Ordovician–Silurian	445–444 Million years ago
2.	Late Devonian	372–359
3.	Permian–Triassic	252
4.	Triassic–Jurassic	201.3
5.	Cretaceous–Paleogene	66
The proposed sixth extinction event		
6.	Holocene	Currently ongoing

The accelerating pattern of revolutions in human history can be viewed as leading to a pulse in this pulsing pattern of mass extinctions. In order to make it into the "Big Six," humans have quite a bit of work to do. This can be accomplished by simply continuing our colonization of the planet for a few million (or perhaps a few hundred) years. A temporary downturn in our present rate of growth would merely delay this process, which would still be consistent with the pulsing pattern. Of course, there are other ways humans can help the Holocene Extinction make the "Big Six." These patterns in Earth history imply that even if under normal circumstances natural or human systems are adapting to some sort of MPP, they can be completely overwhelmed by patterns of the Earth or the Universe that are much larger: asteroids, volcanoes, Milankovitch cycles, or whatever.

Seventy seven years passed between the start of the second and third industrial revolutions. I am writing this sentence seventy seven years after the start of the third industrial revolution. If we view our present situation from a historical point of view—informed by general systems theory—we would have to conclude that another energy revolution is coming. The empirical evidence would lead us to expect the unexpected. What we can say based on the investigation in this book is that the evidence suggests that the evolution of biotic and abiotic systems is guided by the MPP, and this includes human social systems and the moral and religious values associated with them. Consequently, if we find our societies on the way down after the peak, or if the next revolution comes along and propels our societies to new heights, it will have a profound impact on the evolution of the moral and religious values we use to determine what is right and wrong. These values are not based on platonic forms, some static concept of human nature, or an ahistorical definition of the concept of "goodness;" they are the products of an ongoing evolutionary process.

References

References to Nietzsche

A. (1954/1888). The Antichrist [Der Antichrist. Fluch auf das Christenthum]. In *The portable Nietzsche* (W. Kaufmann, Trans.). Viking Penguin.

BGE. (1989/1886). *Beyond Good and Evil* [*Jenseits von Gut und Böse. Vorspiel einer Philosophie der Zukunft*] (W. Kaufmann, Trans.). Vintage.

BT. (1967). *The Birth of Tragedy and The Case of Wagner,* [*Die Geburt der Tragodie*] (W. Kaufmann, Trans.). Random House.

GM. (1967/1887). *On the genealogy of morals* [*Zur Genealogie der Moral. Eine Streitschrift*]. (W. Kaufmann & R. J. Hollingdale, Trans). Vintage.

GS. (1974/1882). *The Gay Science* [Die fröhliche Wissenschaft]. (W. Kaufmann, Trans.). Vintage.

KSA. (1999). *Saemtliche Werke: Kritische Studienausgabe* (2nd ed.). Edited by Giorgio Colli and Mazzino Montinari. Walter de Gruyter.

TI. (1954/1888). Twilight of the idols: Or, how one philosophizes with a hammer [Götzen-Dämmerung. Wie man mit dem Hammer philosophiert]. In *The portable Nietzsche* (W. Kaufmann, Trans.). Viking Penguin.

WP. (1968). The will to power [Der Wille zur Macht]. (W. Kaufmann, Trans.). Vintage.

References to Odum

Odum, H. T. (1977). The ecosystem, energy, and human values. *Zygon., 12*(2), 109–133.

Odum, H. T., & Richardson, J. R. (1981). Power and a pulsing production model. In W. J. Mitsch, R. W. Bosserman, & J. M. Klopatek (Eds.), *Energy and ecological modelling: Developments in environmental modeling 1* (pp. 641–648). Elsevier Scientific Publishing Co. (CFW-81-23).

Odum, H. T. (1982). Pulsing, power and hierarchy. In W. J. Mitsch, R. K. Ragade, R. W. Bosserman, & J. A. Dilon (Eds.), *Energetics and systems* (pp. 33–60). Ann Arbor Science Publishers.

Odum, H. T., & Odum, E. C. (2001). *A prosperous way down.* University Press of Colorado. Kindle version.

Odum, H. T. (2007). *Environment, power, and society for the 21th century.* Columbia University Press.

Other References

Bacon, S. F. (1861). *Works of Bacon.* Brown and Taggard.

Biebuyck, B. (2018). On war and warriors: Friedrich Nietzsche. In D. Praet (Ed.), *Philosophy of war and peace.* Vub Press.

Briggs, J. C. (2017). Emergence of a sixth mass extinction? *Biological Journal of the Linnean Society, 122*(2), 243–248. https://doi.org/10.1093/biolinnean/blx063. ISSN 0024-4066.

Brown, G., & Laland, K. (2006). Niche construction, human behavior, and the adaptive-lag hypothesis. *Evolutionary Anthropology, 15*(1), 95–104.

Brown, J. H., Hall, C. A. S., & Sibly, R. M. (2018). Equal fitness paradigm explained by a trade-off between generation time and energy production rate. *Nature Ecology and Evolution, 2,* 262–268.

Brown, J. H., Birger, R., Hall, C. A. S., & Hou, C. (2024). Life, death and energy: What does nature select? *Ecology Letters* (In Press).

Celente, G. (1997). *Trends: How to prepare for and profit from changes of the 21st century* (p. 337). Warner Books.

Cetron, M., & O'Toole, T. (1982). *Encounters with the future: A forecast of life into the 21st century* (p. 308). McGraw Hill.

Coelho, R. L. (2009). On the concept of energy: History and philosophy of science teaching. *Science & Education, 18,* 961–983.

Cohen, G. A. (2000). *If you're an egalitarian, how come you're so rich.* Harvard University Press.

Bradshaw, C. J. A., & Strona, G. (2022). Coextinctions dominate future vertebrate losses from climate and land use change. *Science Advances, 8*(50), eabn4345. https://doi.org/10.1126/sciadv.abn4345

Cowie, R. H., Bouchet, P., & Fontaine, B. (2022). The sixth mass extinction: Fact, fiction or speculation? *Biological Reviews, 97*(2), 640–663. https://doi.org/10.1111/brv.12816

Curtis, P. (2022). *Nietzsche's will to power: A naturalistic account of metaethics based on evolutionary principles and thermodynamic laws*. PhD diss., Bangor University.

Eldredge, N., & Gould, S. J. (1972). Punctuated equilibria: An alternative to phyletic gradualism. In T. J. M. Schopf (Ed.), *Models in paleobiology* (pp. 82–115). Freeman Cooper.

Eldredge, N., & Gould, S. J. (1993). Punctuated equilibrium comes of age. *Nature, 366*, 223–227.

Emden, C. J. (2014). *Nietzsche's naturalism*. Cambridge University Press. Kindle Edition.

Freud, S. (1930). *Civilization and its discontents* (J. Strachey, Trans.). Hogarth Press.

Fuller, R. B. (1968). An operating manual for space ship Earth. In W. R. Ewald Jr. (Ed.), *Environment and change: The next fifty years*. Indiana University Press.

Glacier, D. S. (2024). Power and efficiency in living systems. *Science, 6*(28).

Hall, C. A. S., & Detwiler, R. P. (1988). Tropical forests and the global carbon cycle. *Science, 239*(4835), 42–47.

Hall, C. A. S. (2017). *Energy return of investment: A unifying principle for biology, economic, and sustainability*. Springer.

Hammond, D. (2003). *The science of synthesis*. University of Colorado Press. ISBN: 9780870817229

Hatab, L. J. (1995). *A Nietzschean defense of democracy*. Open Court.

Henrich, J. (2016). *The secret of our success*. Princeton/Oxford: Princeton University Press.

Hobbes, T. (1668). *Opera philosophica*, Volume III. Leviathan.

Kitcher, P. (1990). Division of cognitive labor. *Journal of Philosophy, 87*(1), 5–22.

Laherrère, J., Hall, C. A. S., & Bentley, R. (2022). How much oil remains for the world to produce? Comparing assessment methods, and separating fact from fiction. *Current Research in Environmental Sustainability, 4*(2022), 100174.

Langbein, W. A., & Leopold, B. (1962). The concept of entropy in landscape evolution. *U. S. Geological Survey Professional Paper 550A*.

Lenton, T. M., Pichler, P., & Weiz, H. (2016). Revolutions in energy input and material cycling in Earth history and human history. *Earth System Dynamics, 7*(2), 353–370.

Lotka, A. (1939). Contact points of population study with related branches of science. *Proceedings of the American Philosophical Society, 80*(4), 601–626.

Lotka, A. (1945). The law of evolution as a maximal principle. *Human Biology, 17*(3), 167–194.

Mayer, J. R. (1978). *Die Mechanik der Wärme: Sämtliche Schriften*. H. P. Münzenmayer e Stadtarchiv Heilbronn Eds. Heilbronn: Stadtarchiv Heilbronn.

Mayer, J. R. (1845). *Die organische Bewegung in ihrem Zusammenhange mit dem Stoffwechsel: Ein Beitrag zur Naturkunde*. Drechsler.

Mohr, S. H., Wang, J., Ellem, G., Ward, J., & Giurco, D. (2015). Projection of world fossil fuels by country. *Fuel, 141*(1), 120–135.

Morrow, D., & Sula, C. A. (2011). Naturalized metaphilosophy. *Synthese, 182*(2), 297–313.

Müller-Lauter, W. (1999). *Nietzsche*. His Philosophy of Contradictions and the Contradictions of His Philosophy, translated from German by David J. Parent. Chicago.

Nagel, T. (1979). *The last word*. Oxford University Press.

Nagel, T. (2012). *Mind and cosmos: Why the materialist neo-darwinian conception of nature is almost certainly false*. Oxford University Press.

Naisbitt, J. (1985). *Megatrends*. Warner Books.

Repetto, R. (1985). *The global possible: Resources, development and the next century* (p. 538). Yale Univ. Press.

Richardson, J. (1996). *Nietzsche's system*. Oxford University Press.

Richardson, J. (2015). Nietzsche's value monism: Saying yes to everything. In M. Dries & P. J. E. Kail (Eds.), *Nietzsche on mind and nature* (pp. 89–119). Oxford University Press.

Sciubba, E. (2011). What did Lotka really say? A critical reassessment of the "maximum power principle". *Ecological Modeling, 222*, 1348.

Simon, J. L., & Kahn, H. (1984). *The resourceful Earth: A response to global 2000* (p. 585). Basic Blackwell.

Steffen, W., Broadgate, W., Deutsch, L., Gaffney, O., & Ludwig, C. (2015). The trajectory of the anthropocene: The great acceleration. *The Anthropocene Review, 2*(1), 81–98.

Sugita, M. (1951). Maximum principle in transient phenomena and its application to biophysics. *Bull. Kobayasi Institute, 1*, 88–101.

Tainter, J. (1988). *The collapse of complex societies*. Cambridge University Press.

Taylor, P. J. (1988). Technocratic Optimism, H. T. Odum, and the Partial Transformation of Ecological Metaphor after World War II. *Journal of the History of Biology, 21*(2), 213–244.

Thimsen, E. (2024). Planetary energy flow and entropy production rate by Earth from 2002 to 2023. *Entropy, 26*(5), 350.

Chapter 7
Conclusion

Abstract This chapter provides a brief review of the material covered in the book and highlights some of the conclusions I have drawn.

The philosophical roots of the maximum power principle (MPP) can be traced back to 1790, when Immanuel Kant published his *Critique of Judgement*: There, he described organic systems as "self-organizing beings" (Kant, 1951, 218–231). We have no evidence that Lotka or Odum were aware of Kant's work, or for that matter Nietzsche's. Kant was influenced by the physiologist Johann Friedrich Blumenbach and his concept of *Bildungstrieb* (formative drive) (1781; Gigantes, 2009, 21). Other philosophers and scientists worked on similar ideas in the nineteenth century. In his development of the concept of the will to power, Friedrich Nietzsche was influenced by the work of Roger Boscovich, Friedrich Lange, Jean-Marie Guyau, William Henry Rolph, Julius Robert Mayer, Carl Nageli, Wilhem Roux, Maximilian Drossbach, and Charles Darwin, among others. There were also a number of scientists working in ecological economics in the nineteenth century that were developing similar ideas.

Nietzsche published most of his ideas about the will to power during the 1880s, the last decade in which he was able to do philosophical work. He thought he had a reason to believe this concept applied to the development of all natural systems, abiotic and biotic. In this investigation, I provided an argument for a unique interpretation of Nietzsche's critique of morality that holds he did not criticize moral values from the usual perspective of some other set of moral values; he described the effects moral values have on the growth and flourishing of life, as outlined by the will to power, understood as an empirical principle. In other words, his "will to power" was not about what "should" be, but rather about what empirically "is." Nietzsche's concept of the will to power and his philosophy were largely ignored during his life. After his death, many aspects of his philosophy went on to have a major impact on philosophy and culture, such as his critique of morality, but the ontological version of the will to power (OWP) was viewed by most philosophers as being empirically implausible.

© The Author(s), under exclusive license to Springer Nature Switzerland AG 2025

T. McWhirter, *Maximum Power and its Philosophical Roots*, SpringerBriefs in Energy, https://doi.org/10.1007/978-3-031-80622-3_7

Twenty two years after Nietzsche's death, in 1922, Alfred Lotka outlined a theoretical argument for his principle of maximum energy flux (PMEF). He argued that those organisms and natural systems that are able to direct more energy through those processes associated with survival and reproduction will have a selective advantage in evolutionary processes. In 1945, he provided an extended argument for this principle and he critically analyzed human behavior from the perspective of it. His work had a large impact on H. T. Odum, who developed Lotka's PMEF further in many different and important ways, and began to refer to it as the maximum power principle (MPP). Working with the physicist Richard Pinkerton, Odum demonstrated the relation between efficiency, rate, and maximum useful power output. Odum went on to describe the relation between the MPP and feedback loops, the pulsing paradigm, and energy quality. He also described how the MPP guided the evolution of moral and religious values and how this evolution is affected by the flow of energy through social systems. He argued that after the approaching peak in our ability to use fossil fuels, moral and religious values will evolve in ways that enable societies to better maximize the power that is available under these new conditions.

There are many parallels between Nietzsche's use of the OWP and Odum's use of the MPP, such as the relation between the tempo of the revaluation of values implicit in Nietzsche's "tension of the spirit" and the pulsing paradigm described by Odum, but the differences between their critical analyses of moral and religious values are particularly informative: among other things, they help us better understand the nature of the perspective they share. Neither analyze moral values critically from the perspective of some ahistorical conception of moral truth; instead, they analyze them from a scientific and empirical perspective, where our understanding evolves over time. By analyzing these differences, as well as the parallels between Nietzsche and Odum, this investigation helps us better understand the "science of morals" that emerges from their work. These differences between Nietzsche and Odum also bring into relief an inconsistency in Nietzsche's work: he describes the evolution of life being guided by the OWP, but he describes the evolution of morality undermining the OWP in his discussion of the "slave rebellion in morality." This inconsistency is resolved in this investigation through a reinterpretation of the value of a mediation between the slave and master moralities. Since the evolution of life is guided by the OWP, the evolution of morality must be guided by it as well, and those moralities that emerge must be products of this evolution, including the 'slave moralities' and the moral values of Christianity.

This investigation has also been able to clear up some confusion in how contemporary scientists have interpreted the MPP. Many have described the MPP as always sacrificing efficiency for maximum power output, but Odum makes it clear that when the available energy is reduced, the optimum efficiency for maximizing useful power output increases. Consequently, Odum argues that the minimum entropy production principle proposed by Ilya Prigogine and J. M. Wiame, which describes the development of systems near equilibrium, is consistent with the MPP (Odum, 1982, p. 35; Prigogine et al., 1946). This demonstrates that the MPP is in a unique position to accurately describe the development of natural systems, both near and far from equilibrium.

This investigation has also been able to illustrate the empirical evidence that has emerged over the past decade and a half that supports the MPP, the PMEF, the OWP, as well as the maximum entropy production principle. Much of this evidence is historical in nature: it illustrates how natural systems develop over long periods of time in ways that increase the energy that flows through them. It therefore lays down a stiff challenge for any principle that is used to describe the evolution of natural systems: it must be able to explain this evidence. When Nietzsche, Lotka, and Odum did their work, they provided arguments for the OWP, the PMEF, and the MPP that were primarily theoretical in nature. At the time, there was not a lot of empirical evidence that was available, but this is no longer the case.

There are some larger philosophical implications of this investigation that are here only briefly sketched. The description of the evolution of moral and religious values through the processes of cultural evolution outlined in this investigation raises questions about the extent to which knowledge and the practice of philosophy itself evolve in a similar fashion. These questions will have to be addressed in a more comprehensive fashion in subsequent work.

Odum's and Nietzsche's view of the evolution of natural systems and moral and religious values can help us better understand the changes that will come after we reach the peak of our ability to use fossil fuels. Their view illustrates how this peak is a unique inflection point in human history with profound implications. This is why Odum took the time at the end of his life to write the book, *A Prosperous Way Down*, which describes the ways that societies will evolve as the energy available to them decreases (2001). This investigation illustrates that human societies and the moral and religious values associated with them have the ability to adapt to new challenges.

The anthropologist Joseph Henrich, who has focused his research on cultural evolution (2016), argues that the secret of our success as a species is our ability to learn and adapt through cultural evolution. One example scientists use to illustrate this ability to adapt is the fact that human beings inhabit many of the different environments around the planet and we have been able to adapt to these different environments relatively quickly. In order for other animals to live in extremely cold environments, they would have to evolve in ways that enabled them to stay warm, e.g., by developing thick fur, through the slow process of biological evolution; humans, on the other hand, simply learned to kill animals that already had this fur and make clothes out of it and they shared this knowledge with other humans. Our ability to learn and share knowledge with others in our time and future generations enables humans to adapt much quicker than other animals.

This investigation has shown that moral and religious values evolve in ways that enable humans to adapt to different conditions. The ancient goal in philosophy was to define, once and for all, the nature of virtue and justice. For many philosophers, the assumption was that there is one set of moral values that are true or the most beneficial. This approach has led many to overlook what Henrich describes as the secret of our success as a species: our ability to adapt through cultural evolution, to change the way we behave, and to change our moral values. This unique ability to adapt will be demonstrated as we deal with the challenges associated with the peak of our ability to use fossil fuels: once again, humans will redefine themselves.

This investigation has attempted to bring different disciplines together in a manner that respects their unique practices, methods, concepts, and styles of writing. This, I would argue, is necessary to fully understand the nature and implications of the work of these two uniquely interdisciplinary thinkers: Howard Odum and Friedrich Nietzsche. Of course, any effort, including this one, will fall short in many respects and scientists and philosophers should be able to extend this analysis further. I welcome this. I hope this investigation serves to encourage more interdisciplinary discussions of the work of Odum and Nietzsche. To help facilitate these discussions, I have created a website—maximumpower.org—where I have posted information about all the references used in this book and I will be posting information about all the work that is done in the future on the MPP, the MEPP, and the will to power. Your comments are welcome.

Nietzsche, Lotka, and Odum believed that human beings were unconsciously fulfilling what these authors considered to be a law of nature. The evidence we reviewed in Chap. 4 supports their view. The thermodynamic interpretation of evolutionary theory discussed in this investigation describes our bodies and minds evolving in ways that enable social systems to maximize useful power output: our ability to see, hear, feel, walk, run, sit, eat, sleep, our feelings, desires, and fears, our ability to remember or forget things, our ability to work with others, your ability to read this sentence, are all products of this evolutionary process, as well as the music and art that moves us, the moral and religious values we follow, and the religions, philosophies, and scientific theories we believe to be true.

References

Blumenbach, J. F.. (1781). *On the drive for education and the business of procreation* (1st ed.). Johann Christian Dieterich, Göttingen (online).
Gigantes, D. (2009). *Life: Organic form and romanticism*. Yale University Press.
Henrich, J. (2016). *The secret of our success*. Princeton University Press.
Kant, I. (1951). *Critique of judgment* (Bernard, Trans.). Hafner Press. Kritik der Urteilskraft.
Odum, H. T. (1982). Pulsing, power and hierarchy. In W. J. Mitsch, R. K. Ragade, R. W. Bosserman, & J. A. Dillon (Eds.), *Energetics and systems*. Wiley.
Odum, H. T., & Odum, E. C. (2001). *A prosperous way down*. University Press of Colorado. Kindle version.
Prigogine, I., & Wiame, J. M. (1946). Biologie et Thermodynamique des Phenomenes Irreversibles. *Exerientia, 2*, 451–453.